SHIPPING POLICIES FOR AN OPEN WORLD ECONOMY

SHIPPING POLICIES
for an open world economy

EDITED BY G.N. YANNOPOULOS

ROUTLEDGE
London and New York

First published 1989
by Routledge
11 New Fetter Lane, London EC4P 4EE
29 West 35th Street, New York, NY 10001

Printed and bound in Great Britain by
Biddles Ltd, Guildford and King's Lynn

British Library Cataloguing in Publication Data

Shipping policies for an open world economy.
1. Shipping services. Policies of governments
I. Yannopoulos, George (George Nicholas), *1936–*
387.5′4

ISBN 0-415-00400-4

Library of Congress Cataloging-in-Publication Data

Shipping policies for an open world economy / edited by G.N. Yannopoulos.

p. c.m.
Includes index.
ISBN 0-415-00400-4
1. Shipping. 2. Merchant marine. 3. Shipping bounties and subsidies. 4. Competition, International. I. Yannopoulos, George N., 1936–
HE581.S47 1989
387.5′.068–dc19 88–30771
CIP

Contents

List of Tables

Abbreviations

ACP	African, Caribbean and Pacific states of the Lomé Convention
ASEAN	Association of South-East Asian Nations
BMCF	British Maritime Charitable Foundation
CAACE	Association of European Community Shipowners
CAP	Common Agricultural Policy
COMECON	Council of Mutual Economic Assistance
CSG	Consultative Shipping Group
EEC	European Economic Communities
FEFC	Far East Freight Conference
FMC	Federal Maritime Commission
FOC	Flags of Convenience
GATT	General Agreement on Tariffs and Trade
GCBS	General Council of British Shipping
LDCs	Less Developed Countries
NATO	North Atlantic Treaty Organisation
NIMO	New International Maritime Order
OECD	Organisation for Economic Cooperation and Development
POL	Polish Ocean Liner
TMC	Traditional Maritime Countries
UN	United Nations

UNCTAD	United Nations Conference on Trade and Development
UNICE	Association of European Community Industrialists
UWIST	University of Wales Institute of Science and Technology
WITASS	Association of West India Transatlantic Steamship Lines.

Preface

This book brings together the contributions to the conference on the International Maritime Transport Policy of the European Communities held under the aegis of the Graduate School of European and International Studies in Wiston House, Wilton Park Conferences, Steyning in the spring of 1986. The papers have been updated in the light of subsequent developments and in particular the endorsement by the Council of Ministers of the EEC of four of the Commission's agreed policy proposals in December 1986. These contributions examine the implications for the international environment in which world shipping operates of the new Common Maritime Transport Policy of the EEC and of policy developments in other traditional maritime transport countries. The links between these developments and the GATT initiative on international trade in services are also explored.

The conference was made possible through the financial support provided by Mr S Clinton-Davis, the Commissioner for Transport of the Commission of the European Communities and Mr J Ad Hadjipateras, Chairman of the London based Greek Shipping Cooperation Committee. Support and encouragement was also provided by the then Undersecretary responsible for Shipping in the Department of Transport in HM Government, The Rt. Hon. The Earl of Caithness who opened the proceedings of the conference. Mr G R Denton, Director of Wilton Park Conferences ensured that the best of the facilities of Wiston House were made available to the conference participants. I am most grateful to all of them for their support.

My thanks also go to Mrs Ann Bedford, the Secretary of the Graduate School of European and International Studies, who assisted me in the preparation of the conference and to Mrs Margaret Paul who typed the manuscript most skillfully and speedily.

G N YANNOPOULOS
Reading, March 1988

Notes on the Contributors

Hans Böhme is Research Fellow at the Institut für Weltwirtschaft of the University of Kiel, Federal Republic of Germany.

Mark C Casson is Professor of Economics and Head of the Department of Economics at the University of Reading, England.

Patrick Gautrat is General Director of the Central Committee of the French Shipowners, Paris, France.

Richard Goss is Professor of Maritime Economics and Administration at the University of Wales Institute of Science and Technology, Cardiff, Wales.

Raymond J Krommenacker is Counsellor at the Secretariat of the General Agreement on Tariffs and Trade, Geneva, Switzerland.

Gerald Seifert is General Counsel for Maritime Policy of the Committee on Merchant Marine and Fisheries of the US House of Representatives, Washington, DC, USA.

Arnljot Strømme Svendsen is Professor of Maritime Economics and Director of the Institute for Shipping Research, Bergen, Norway.

John Tzoannos is Professor at the Athens School of Economics and Business Science, Athens, Greece.

George N Yannopoulos is Chairman of the Graduate School of European and International Studies, University of Reading, England.

CHAPTER 1

WORLD SHIPPING BETWEEN LIBERALISM AND PROTECTIONISM : AN INTRODUCTION

George N Yannopoulos

1 Return to a market-based organisation

Interest within the OECD area in the liberalisation of trade in services has grown considerably since the completion of the Tokyo Round of Multilateral Trade Negotiations. Once the GATT rounds brought tariff barriers to merchandise trade to low levels, further benefits in stimulating world trade through competition can be derived either (a) through the elimination of non-tariff barriers, or (b) by extending the process of liberalisation to world trade in services, or (c) through a combination of both approaches (a and b).

World trade in services has grown in importance and has been expanding at rates comparable to the growth of world merchandise trade. More significantly exports of services appear to be a correlate of economic growth. Work by Riddle (1987) shows that for the lower-middle-income countries in particular, high growth tends to be associated with a significantly higher proportion of services in total exports.

However, the GATT framework cannot be used in its current form to bring forward a more open regime in the international trade in services since the General Agreement does not provide explicitly and directly a legal basis to deal with the issue. In an important document (TC/WP85/9 16 December 1985) the OECD proposed a set of general principles under the title *The Elements of a Conceptual Framework for Trade in Services*. Following the discussions in Punta del

Este in September 1986 for a new GATT round the international trading regime in services has become a dominant theme in international trade policy fora.

The case for policy reform on questions relating to international trade in services has been further strengthened by the deregulation movement that influenced economic policy in several OECD countries since the second half of the 1970s (Backman, 1981).

Another development with important implications for the future of the international regime of trade in services is the fact that in several regional trading blocs attempts have been made to extend the process of economic integration to fully cover trade in services. The most significant of these attempts are the initiatives of the European Community to work out a common transport policy, to harmonise banking and insurance rules and to introduce a new common policy on tourism. These regional liberalisation attempts acquire particular significance in view of the existing institutional vacuum in the international trade regime. Regional integration *per se* does not automatically lead to freer world trade. The analysis of the international implications of the regional integration initiatives thus acquires considerable importance.

The shifts in policy positions described above have exerted an influence on almost all sectors of world trade in services. International trade in maritime transport services is no exception. Indeed, in some respects it has found itself in the forefront of the discussions for worldwide liberalisation of trade in services. New rules governing the conduct of international maritime practices were introduced in the United States through the Shipping Act of 1984 (see chapter six of this volume). In the European Community and after a long period of benign neglect, the Council of Ministers adopted in December 1986 a set of four regulations that underline the European consensus in favour of a multilateral and market-based organisation of world shipping (see chapter three of this volume). In a related development, some Governments (France, Norway, the United Kingdom etc.) have allowed the formation of new international registries (Kerguelen Island, the NIS, and the Isle of Man respectively) linked to the national flag and acting in practice as 'captive open registries'. Instead of controlling the process of flagging-out, governments are now ready progressively to equalise the conditions between closed and open registry systems (Kappel, 1987).

It is sometimes said that the international maritime transport sector has traditionally enjoyed a comparatively more liberal regime (Richardson, 1987). This may be true for some segments of the sector but does not necessarily hold true for the whole of the market. The defence role of the merchant fleet in providing transport supplies during periods of military conflict (Stryker, 1986) and its strategic value because of its links with shipbuilding have provided excuses for protection. However, in most countries the protectionist measures introduced in order to serve defence objectives have in fact little 'defence content'. Furthermore, by crippling the competitiveness of the national fleet such measures lead to its shrinkage thus undermining the original defence objectives. The proliferation of government induced restraints to achieve defence and industrial policy objectives has progressively eroded the open nature of the earlier regime. Cargo sharing, cargo preferences, cabotage reservations and bilateral agreements have become widespread practices (see chapter two in this volume). The costs from such policies are estimated to involve, only in the United States, a little under $3,000 million per annum in terms of transfer from taxpayers and from shippers (White 1987). The inefficiency costs are unlikely to be far below. Another factor that has undermined the open character of the international maritime transport market has been the emergence of state shipping companies. The traditional liberal international regime in maritime transport came further under attack during the heydays of the movement for a New International Economic Order. A number of developing countries used UNCTAD, the specialised agency of the United Nations on trade and development issues, to push for redistributive measures aiming at the strengthening of their own merchant fleets through internationally sanctioned cargo reservation rules. This demand led eventually to the UN Code for Liner Conferences which reserves cargo among the exporting country, the importing nation and cross-traders on the basis of the formula 40:40:20. The success of UNCTAD in regulating competition in the liner conferences segment of the world maritime transport industry prompted subsequent moves for intervention in the fairly unregulated open registry system in order to reduce the advantages of the flags of convenience. This move came to a halt because the search for a New International Economic Order that propagated it has lost its momentum in the meantime. Indeed the phase of regulatory fervour in international maritime transport has been remarkably short-lived. A new era of worldwide liberalisation seems to be on the ascent. The US Shipping Act 1984, the European Community's Common Maritime Transport Policy and the emergence of the 'captive open registries' in some traditional maritime nations are clear indications of this. The shift from government controlled to market-based shipping policies can be understood within the context of the major changes that have taken place in the international economic environment. In a period of slow-growth the intensification of competition in the world markets makes companies more conscious of the need for cost effectiveness. Protectionist measures in the services sectors have "spillover" effects on those industries for which they provide inputs. The costs of interventionist measures in the international markets for services are borne not only by the directly affected sector but also by those industries outside it that use such services as inputs. A competitive manufacturing industry requires a flexible and open international shipping system. Governments have thus become more aware of the consequences of departures from market -based policies in international shipping (Schott and Mazza, 1986). Secondly, the shipping industry is faced with extensive adjustment problems especially in the traditional maritime countries as the ease for newcomers to enter the trade is facilitated during a slump in world demand when the selling of ships increases and the ship lending policies of the financial institutions change. The need for reducing government imposed costs on the shipping enterprise becomes more pressing in order to facilitate adjustment. Thirdly, the rate of technological innovation in the shipping industry has also increased. Containerisation, new opportunities for inter-modal co-ordination and increased use of telecommunications and computers in shipping operations offer new sources of comparative advantage in international maritime transport. The spearheading of these technological advances for the purpose of sustaining comparative advantages requires a more competitive environment - not the stifling framework of government restraints.

2 A European framework for international shipping

Of the various recent policy developments, the emergence of a common maritime transport policy in the European Community is undoubtedly a major factor that will crucially influence the future design of the international maritime order. The EEC merchant fleets account for about one-fifth of world tonnage (see table 1.1). If one adds the fleets in open registries owned by Community nationals then the share of world shipping controlled by Community nationals or Community based companies goes up by about seven per cent.

Table 1.1: European Community Merchant Fleets

(million tons gross)

	1975	1983	1986
EEC (12)	103.2	104.4	77.3
Percentage of World total	(30.2)	(24.7)	(19.1)
Belgium	1.4	2.3	2.4
Denmark	4.5	5.1	4.7
France	10.7	9.9	5.9
Germany	8.5	6.9	5.6
Greece	22.5	37.5	28.4
Italy	10.1	10.0	7.9
Netherlands	5.7	4.9	4.3
Portugal	1.2	1.2	1.1
Spain	5.4	7.5	5.4
United Kingdom	33.2	19.1	11.6

Source: ISL, Bremen

However, the new common maritime transport policy (now embodied in Regulations Nos. 4055, 4056, 4057 and 4058/86) will affect all shipping concerns engaged in the transportation of the Community's seaborne trade irrespective of flag. This is because the EEC common maritime transport policy does not limit itself to internal harmonisation but, significantly, introduces global regulations for extra-Community trade in maritime transport services. As the Community is the world's largest trading area the impact of the EEC common maritime transport policy will extend over a considerable part of the international market for maritime transport. In particular, the EEC accounts for 21% of world imports and 20% of world exports by value compared to 16% and 15% respectively for the USA and 9% and 10% respectively for Japan. More relevant is the fact that about 95% of the value of Community trade with third countries (non-members) is carried by sea (the corresponding figure for intra-Community trade is 30%). Out of this seaborne trade approximately 60% is carried in ships flying flags of non-member states. Part of it, however, involves open registry shipping. The Commission services estimate that ships under flags of convenience owned by Community nationals account for approximately 23% of the cargo generated by extra-area merchandise trade. This leaves to third country shipping firms a little less than 40% of the tonnage of Community trade with non-members.

Community policies are not the only ones that influence developments in European shipping. The net earnings of Community shipping are derived by about 50% from cross-trading outside the Community. There exists therefore a strong interdependence between the Community and the other maritime nations which has important implications for policy coordination. As the existing institutional framework does not provide an appropriate basis for such coordination, it is important - in view of this interdependence - that the EEC possesses the appropriate policy instruments to respond to ad-hoc maritime policy initiatives of other nations. Indeed, among the aims of the common maritime transport policy is the provision of such policy instruments. On the basis of the new EEC policy, the European Community can initiate co-ordinated action if free access to cargo is restricted by impediments imposed by third countries. Furthermore, when third country shipowners in liner shipping engage in unfair pricing practices causing injury to Community shipowners redressive duties may be introduced by the EEC. This extension of the anti-dumping policy to shipping has however some dangers of degenerating into a protectionist device. It will be mistaken to deduce from the above that the common maritime transport policy is essentially co-ordinated retaliatory action to counter protectionist policies of other countries and restrictive business practices (For more details, see Chapter 3). The policy is fairly extensive although not comprehensive. It envisages the eventual phasing out of all existing bilateral cargo sharing agreements involving EEC countries and the prohibition of new agreements. Restrictions on freedom to provide maritime transport services between EEC countries and third countries are also to be phased out. And finally, it provides detailed rules for the application of articles 85 and 86 of the Treaty of Rome to the maritime transport sector which enable liner conference agreements to go ahead, but simultaneously give the shippers the right of complaint about anti-competitive practices. It appears that although internal harmonisation does not lack in detail, nevertheless it falls short of being comprehensive. Removing intra-EEC cabotage restrictions has not proved possible and the Council of Ministers in their December 1986 meeting referred the issue back for reconsideration by a special working group. This suggests that progress on this issue is unlikely to be rapid.

3 Plan of the book

This volume gathers together eight especially commissioned papers whose aim is to appraise the international maritime transport policies and developments in the European Community, the United States, the non-EEC Scandinavian countries and the state trading countries in the light of the GATT initiative to liberalise world trade in services. The contributors to this volume examine the implications of several policy actions and particularly of the establishment of the EEC's common policy on the maintenance of free and fair competition in world shipping markets and enquire as to their impact on the competitive position of different shipping operators.

In order to place the subsequent discussion and evaluation within a specific context, the paper by Dr Hans Böhme (Chapter 2) analyses in a systematic way the economic consequences of three types of government-induced distortions: state subsidies, flag discrimination and price (freight rate) controls. This chapter provides a useful framework for the analysis that follows in the subsequent chapters of the volume. Chapter 3 presents a detailed summary of the EEC's common maritime transport policy and discusses its relation to the various attempts to liberalise world shipping markets. The chapter traces the history of the formulation of the original Commission proposals, illustrates the conflicting interests of the member-states of the EEC and summarises the essential policy decisions taken by the Council of Ministers in December 1986. Professor J Tzoannos also makes a short comparison of the original Commission proposals with the provisions finally adopted in the relevant Regulations. One important difference between the original proposals and the compromise solutions finally endorsed by the Council of Ministers is the failure to agree on intra-EEC cabotage trades. Some member-states who have had restrictions for many years on cabotage trade were loath to remove them, making impossible the emergence of a consensus on the matter. It became clear that the political muscle of some highly protected shipping groups has influenced their governments' political will to achieve a consistent policy. For there may exist policy linkages here. The longer this protectionist policy stays the more cross-trading opportunities will be lost and the more other countries will wish to implement increasingly protectionist and restrictive regimes. Given these policy linkages it is clear that the Community would have achieved a more decisive impact in reshaping the international maritime transport regime if it decided as a matter of priority to put its own house in order by removing at an early stage the restrictive laws which deny shipowners from all member-states of the European Community the opportunity to compete for trade on an equal footing.

The author of Chapter 3 also takes a critical stand on the measures for co-ordinated action against other countries preventing EEC merchant fleets from having access to cargoes or permitting their shipping companies to indulge in predatory pricing. He fears that such action may invoke retaliatory response by other shipping nations particularly if the latter see it as a covert protectionist device. However, it is worth recalling here that analysts of merchandise trade policy have argued (Goldstein and Krasner, 1984) that a policy of appropriate unilateral trade retaliation is a more efficient way to deal with unfair trade practices. Such a policy when exercised with care may in practice elicit co-operation and freer trade rather than lead to trade wars.

A further critical appraisal of the common maritime policy of the EEC is presented in the following chapter by Professor R Goss. Of the many interesting issues raised in Chapter 4 two deserve to be singled out: the efficiency implications of piecemeal liberalisation and the relevance of the common policy in improving the competitiveness of European shipping.

Liberalisation on a sectoral basis may not be an optimal policy if other economic sectors in the European Community (e.g. agriculture) are heavily protected and sheltered from world competition. Exposing one sector to world-wide competition without at the same time removing the burdens imposed on it from the protection of other sectors, may actually undermine its ability for competitive survival even if that sector enjoys comparative advantage in a fully liberalised regime. Work by A Stoekel (1985) on the sectoral repercussions of the European Community's Common Agricultural Policy supports Professor Goss' view. Simulations using a general equilibrium framework showed that in the absence of agricultural support by the EEC, Community exports of all services would have been higher by 17 per cent and the imports of all services lower by 6 per cent. The opportunity of the Uruguay Round which brings together issues of agricultural trade policy and trade in services is too important to be missed by the Community. The other question raised by Professor Goss in his contribution to this volume relates to the limitations of the common policy in tackling problems of competitiveness of European shipping. In an international market where both national and convenience flags co-exist, shipping companies operating under the flags of the traditional maritime nations are forced to absorb non-wage crew costs. This backward shifting on to the employer of the burden of non-wage crew costs (i.e. social security and seamen's pension schemes) has been singled out as the most important factor contributing to the loss of comparative advantage of European shipping. In this respect, complementary policy measures are required. Issues of competitiveness in shipping are also discussed in Chapter 5. This chapter based on the findings of the New Industrial Economics literature provides interesting suggestions for the competitive survival of the shipping enterprise of the traditional maritime nations. Higher quality of service as a market strategy to overcome the trend towards a loss of comparative advantage in the traditional areas of shipping activity is emphasised in this chapter by

Professor M C Casson. Professor Casson makes an interesting distinction between production quality and transactions quality and points out that it is feasible to work out alternative institutional arrangements in maritime transport markets that will enable Community shipping to move upmarket in terms of better quality of service. This is in contrast to Commission views (Commission of the EEC, 1985) according to which the advantages to be derived from higher quality of service "are becoming increasingly costly to achieve". The following three chapters (6 to 8) examine the maritime transport policies of three major actors in world shipping: the United States, the non-EEC Scandinavian countries and the Eastern European countries. The aim of this comparative exercise is to show the extent to which these policies move in a parallel way with the EEC's common maritime transport policy towards world-wide liberalisation and the areas where they create restraints and harbour protectionism.

G Seifert, in Chapter 6, stresses the strategic thinking (essentially a high degree of shipyard mobilisation) influencing US maritime policy and predicts that such considerations are unlikely to help towards the repeal of the Jones Act and the removal of existing cabotage reservations and flag preference on agricultural donations cargoes. However, the author emphasises the essentially open market orientation of the US Shipping Act of 1984 which brought about the liberalisation of liner markets. There are several similarities, the author of Chapter 6 points out, between the US Act of 1984 and the EEC common maritime policy adopted in December 1986. Nevertheless, a concern is expressed about the danger of using some of the provisions of the new EEC Regulations for introducing countervailing measures against US carriers.

The issue of maritime relations with the state trading companies is raised by P Gautrat in Chapter 7. The author regards the availability of countervailing measures in the new Regulations of the common maritime transport policy as useful Community policy instruments to ensure through negotiations the normalisation of the EEC's maritime relations with the USSR and its allies and the elimination of the disruptive effects of their policies. However, the author doubts the feasibility of structured negotiations on the basis of a Community-wide mandate given the special political or commercial interests of some EEC states.

Professor Svendsen analyses in Chapter 8 the changing strategies of the two non-EEC Scandinavian countries, Norway and Sweden. The author emphasises that under the impact of the recent crisis the non-EEC Scandinavian shipping industry has been restructuring by moving away from traditional shipping to maritime intelligence activities (management, brokerage etc.), offshore operations and aquaculture. This marketing strategy, the author believes, will make the maritime industry of these countries more complementary rather than more competitive to the EEC shipping industry. The last chapter explores the issues that are likely to be raised during the Uruguay Round if shipping is included as a priority sector in the negotiations for the liberalisation of trade in services. There are special problems in deciding on the conditions that will ensure unrestricted market access because some commercial presence of the producer of the services is required in the market of the buyer. The application of some other GATT principles that have helped decisively in the past the process of liberalisation of world merchandise trade (transparency, most favoured nation treatment, etc.) raises some interesting questions because of the special characteristics of the trade in services. These issues are discussed in the contribution of Dr R Krommenacker in the concluding chapter of this book.

4 Liberalisation strategies

The policy recommendations included in the eight chapters of the present volume and the discussions during the deliberations that took place at the conference where the eight contributions were first presented, have sketched a number of alternative options that can be used to steer world shipping towards an open market. These options can be described as the minimalist approach, the incrementalist option, the public choice alternative and the maximalist scenario.

The *minimalist* solution, appearing in the chapter by H Böhme, aims at stabilising the *status quo* of developing countries' trades thus checking the tendency for the restraints to spread further. An integral part of this minimalist strategy is the adoption of a common position by all traditional maritime countries against attempts to extend cargo sharing and other forms of discrimination from liner shipping to other segments of the maritime transport industry. To strengthen the bargaining ability of the traditional maritime nations to press effectively for a more liberal world maritime order, the minimalist approach recommends that European Community countries should extend and solidify freedom to supply shipping services in their mutual traffic by removing restraints such as cabotage reservations, cargo preferences and bilateral agreements.

The *incrementalist* approach has been advocated by G Seifert. According to this 'step by step approach' the obvious next opportunity will be the forthcoming review of the United Nations Code of Conduct on Liner Conferences. A common negotiating strategy by traditional maritime nations can force a change from the 40:40:20 cargo reservation formula to a 33:33:33 arrangement. Indeed, the United States is expected to adopt this alternative formula in their bilateral negotiations with China and the USSR.

Another alternative towards progressive liberalisation is the *public choice* option - originally suggested by Professor M C Casson during the discussions in the conference from the proceedings of which this volume is derived. Professor M C Casson made a distinction between property rights and the use of these property rights. The UN Code sacrifices efficiency in the supply of shipping services in favour of short-term inter-country redistribution of income from shipping. This conflict between allocative efficiency and income redistribution could be resolved if the right to share cargo given to countries is subsequently offered through auctioning of the corresponding licences under competitive conditions open to both national and international shipping companies. This arrangement will enable the governments of the developing countries to obtain a share in shipping income without affecting adversely the efficiency of the world shipping industry. The competitive auctioning technique can also be used to allocate among European shipping companies the 50 per cent share to be available to Community fleets when the present conditions of bilateralism with the Comecon countries are replaced by a Community-wide framework.

The *maximalist* approach would extend the competence of GATT to trade in services and apply the multilateral negotiations technique to this type of trade. GATT has succeeded in considerably liberalising world trade in manufactures during the past forty years. The minimum objective at the first stage of the GATT involvement in maritime services is to establish an equivalent to the MFN principle in merchandise trade, to trade in maritime transport services. The next step will be to prepare a framework agreement under the GATT to ensure free market access in the provision of maritime transport services. The technical difficulties in achieving an enforceable multilateral agreement are not exhausted by these two issues. Questions of transparency, regulatory policies and safeguards also require solutions as does the desire to maintain a balance of rights and obligations for the participants. If such difficulties prove insurmountable in the GATT fora, then the maximalist scenario has a fall-back position consisting of joint action within the OECD area to create a broader regional free market in maritime transport services among all countries willing to adopt a liberal maritime order.

REFERENCES

Backman J (1981) The Problem of Regulation in J Backman (ed) *Regulation and Deregulation*, Bobbs-Merrill Educational Publishing, Indianapolis, pp 1-34.

Commission of the EEC (1985) Progress Towards a Common Transport Policy: Maritime Transport, *Bulletin of the European Communities*, Supplement 5.

Goldstein J L and Krasner S D (1984) Unfair Trade Practices: The Case for a Differential Response, *American Economic Review*, Vol 74, No. 2, May, pp 282-7.

Kappel R (1987) New Horizons in the Internationalisation Process of Shipping, *ISL* , March, pp 4-5.

Richardson J (1987) A Sub-sectoral Approach to Services' Trade Theory, in O Giarini (ed) *The Emerging Service Economy*, Pergamon Press, Oxford, pp 59-82.

Riddle, D I (1987) The Role of the Service Sector in Economic Development: Similarities and Differences by Development Category in O Giarini (ed) *The Emerging Service Economy*, Pergamon Press, Oxford, pp 83-104.

Schott J J and J Mazza (1986) Trade in Services and Developing Countries, *Journal of World Trade Law*, Vol 20, No. 3, pp 253-73.

Stoeckel A (1985) *Intersectoral Effects of the CAP: Growth, Trade and Unemployment*, Occasional Paper No. 95, Bureau of Agricultural Economics, Australian Government Publishing Service, Canberra.

Stryker R F (1986) Civil Shipping Support for NATO. *NATO Review*, No. 1, February, pp 29-33.

White L J (1987) International Trade in Ocean Shipping Services: The United States and the World, in Trade Policy Research Centre, *Trade in Services, Open Markets and the Uruguay Round Negotiations*, Conference Report, London 8 July, pp 227-47.

CHAPTER 2

THE ECONOMIC CONSEQUENCES OF RESTRAINTS ON TRANSACTIONS IN THE INTERNATIONAL MARKET FOR SHIPPING SERVICES

Hans Böhme

1 The traditional and the new maritime order

For almost one hundred years - from the repeal of the British Navigation Acts in 1850 to the early post-World War II years - ocean shipping enjoyed the most liberal framework for its operations that could be conceived. With very few exceptions,[1] national and foreign vessels were treated equally, access to the shipping markets was free for shipowners and shippers alike, and on the whole, shipping services could be produced with a maximum degree of efficiency. This framework corresponded to the theory of international trade policy, creating a division of labour according to comparative advantage and pledging non-interference by public authority. There were a few black spots, such as the reservation of national coastal traffic (cabotage) for national-flag vessels, or the payment of subsidies (though not on a large scale at least before 1914). Shipping conferences which apply certain restrictions to the markets were permitted on the assumption - corroborated by repeated investigation into their operation and market behaviour - that they offer the best conditions for efficient, cheap, and stable liner services, with reasonable freight-rates and required frequency of service. While loyalty ties were permitted, and the market shares of conferences tended to be high, there were no administrative barriers to the entry of newcomers into the markets. The terms of shipment were left to commercial negotiation.

The traditional *Freedom of the Seas* order has steadily crumbled away since about 1950. Subsidies have greatly increased, though they often support national shipbuilding industries rather than shipping. Other forms of interference with the market mechanism have been introduced, known under the catchword of "flag-discrimination". The United States had set a precedent by issuing cargo preference legislation in the early 1950s. [2] Subsequently, the leading role towards what is now termed the *New International Maritime Order* (NIMO) was quickly taken on first by individual developing countries, and later by the UNCTAD division of shipping. In the context of the general economic and political aims of promoting the developing economies by turning away from the traditional liberal framework, new concepts were developed. [3] These concepts incorporate such notions as "a country's own foreign trade", "its right to carry its oceanborne trade", "national carrier", or "liner shipping as a public utility". Efficiency, profitability, comparative advantage and related criteria were to a large extent displaced by aims of equity, income distribution, etc. The traditional economic criteria have lost ground in shipping as well as elsewhere. Discrimination in shipping was re-defined.

Meanwhile, a quadrangle of shipping policy has developed. The traditional philosophy of the *Freedom of the Seas* is being upheld only by the so-called Traditional Maritime Countries (TMC) of Europe, partly also by the United States which has recently taken some steps away from its long-standing protectionist policy, and by some overseas countries. Most developing countries are in favour of a closely controlled framework. They rely much less on the market mechanism than on public authority. This implies that instruments are being applied on a large scale which are considered to be restraints on the market from the viewpoint of the TMC. Finally, there are the countries of the Eastern bloc where shipping is state-owned as part of the all-out system of state-economy.

The problem of restraints in shipping, and of intervention with competition and the market mechanism, has thus continually gained in importance. Added to public restraints is a tendency to view the operation of liner conferences and other types of "commercial practices" more critically. [4] It is true that the criticism which had forcefully been raised by LDCs during the 1960s, has largely died away since national-flag carriers of many LDCs have become conference members. However, certain changes are taking place in liner shipping technology and organisation which are either claimed to be hampered by conferences in their present form, or which are believed to enhance the market power of conferences.

In the more recent past, international shipping policy has come under scrutiny in the context of various new trends. These are, first, the drive towards liberalisation of trade in services, in the framework of GATT negotiations. Second, the attempt to establish a common West European shipping policy, with the European Commission in charge. [5] This includes a reconsidered role of liner organisation and appropriate legal basis for conference operations. Triggered largely by the problems which the imminent ratification of the UN Liner Code created in 1975, a more full-fledged shipping policy approach has developed which led to the March 1985 Commission proposal on shipping policy. Third, the United States, which has been opposed to the UN Liner Code from its inception, has held talks with CSG countries on mutual guarantees for open access to liner markets. [6] She has also taken steps to abolish restrictive or promotional elements of her own shipping policy. There are, moreover, the UNCTAD negotiations on such issues as the Flags of Convenience, or a Bulk Cargo Sharing Code. A definite clash of interests is visible between the more traditional maritime countries, and certain developing countries, especially in South America and West Africa. The UN Liner Code is playing a prominent role in such discussions (though its actual role as a source of protectionism is often overvalued).

It is the object of this chapter to discuss the impact of restraints on competition in ocean shipping markets, especially to give an overview of their more important effects on economic efficiency in shipping. It will not attempt to assess their effects on the (national or world) economy at large. Likewise, it will not be possible to quantify such effects. The chapter focuses on *public* restraints. So-called private restraints or commercial practices, mainly by liner conferences, will not be considered. Conferences are still regarded as the appropriate form to organise rational liner services; their operation must largely be evaluated from this angle.

Conference practices would simply serve to protect their members only if it could be proven that the properties associated with conferences, and held to be vital to efficient regular shipping services, were in effect also part of individual liner services operating without any conference framework. It should be noted, however, that conferences are accepted in Europe on the understanding that market access must be open for all. In addition, there is evidence that the empirical relevance of conference practices is diminishing under the impact of new trends in technology and organisation of liner shipping. [7] There are, however, certain problems caused by regulatory policies.

2 A short review of relevant measures and procedures

Though the chapter is not meant to present a taxonomy of restraints, it seems appropriate to mention the most important categories, and to recall their political or economic rationale. The aim is usually to ensure participation of a country's fleet in ocean shipping, or to establish such a fleet in the first place, and then to promote its development. There is, accordingly , a desire to alter the division of labour in shipping as it would evolve from the unfettered operation of market forces, on the basis of comparative advantage. Without protection, the national fleet would either be non-existent, or smaller, or differently structured, because of a lack of competitiveness due to high costs, lack of quality, or some similar reason. Though such policies are often based upon infant-industry considerations, pertinent measures are usually more persistent and not only temporary.

There are three main types of restrictive or promotional measures which are applied to attain such basic goals. These types are:

- the payment of subsidies,
- instruments of flag-discrimination, and
- control and/or setting of freight rates, for example with the object of securing low rates for a country's exports.

Most important is flag-discrimination, that is, measures which grant preferential treatment to national vessels, or impose disadvantage upon foreign-flag ships. [8] This is obviously in contrast to the principle of equal treatment of all ships regardless of flag, according to a general inland-parity status. Flag-discriminatory measures are directed at the demand for shipping services by influencing the cargo flows, as well as at the supply of ships by influencing their availability in kind, or their competitiveness (via their costs).

Instruments of shipping policy which involve flag-discrimination are manifold [9]; they include:

- reservation of part of or of all cargo moving in a country's foreign trade, sometimes on the basis of assumed "property right in its own national foreign trade"; this is flag-discrimination proper;

- special levies on foreign-flag ships, or preferential dues and levies for national flag ships (for example, lighthouse dues, pilotage fees, part dues, etc, passage monies for using the territorial waters, however, being forbidden by the international law of the sea).

- preferential treatment granted to national-flag ships in national sea-ports, priority allocation of berths, and so on;

- the use of the terms of shipment as a device to influence the choice of vessel by shippers;

- the establishment of cargo-booking monopolies (government or semi-government), as a means to control and allocate cargo flows, or the admission of only certain "designated carriers" to a country's shipping markets.

Such measures, especially if they involve cargo reservation or allocation, may be taken unilaterally, or bilaterally by trading partners, or multilaterally, for example by international convention. Cargo may be completely reserved, or reserved to some percentage, in favour of one country, or split up between two, or allocated to flags including third-flag carriers (cross traders). Its most stringent form is the 50:50 clause which is tantamount to complete autarchy in shipping services.[10]

The payment of subsidies is relevant mostly as far as operating subsidies are concerned which have a direct bearing upon the competitiveness of national ships. While they clearly distort the division of labour away from the pattern determined by comparative advantage, they are not restraints in the proper sense of barriers to access, or some other form of not "market conforming" measures.

3 Effects of restraints on shipping efficiency

In the following section, the effects of the three main types of restrictive government policy mentioned above will be studied and their impact on the efficiency of the ocean shipping system will be examined.

Evaluating the effects of restraints involves two stages, first, analysis of empirically observed or theoretically derived effects, and second, the assessment of their actual impact on shipping. In the second stage, several mitigating factors, limitation to certain areas of the shipping market, and long-term market trends must be considered. There may be very considerable differences between potential, and actual effects, and these may increase or decrease over time.

Restraints may affect numerous aspects of shipping operations, such as freight rates, costs of services, fleet sizes, fleet structure, demand for shipping services and freedom of shippers' choice of carrier, ship operation and organisation of services, employment pattern and opportunities for ships, and in general terms, efficiency of shipping services.

Flag discrimination restricts access to cargo for shipowners and the right of choice of vessel for shippers; it interferes with the freedom of access to waterways and sea-ports on the same terms for all ships regardless of flag; and it discriminates between national and foreign-flag vessels with respect to the use of port and harbour facilities.

Subsidies in favour of national-flag ships, on the other hand, do not interfere with the day-to-day operation of ships, and do not interfere directly with such aspects of competition as the acquisition of cargo, or the right of shippers to choose a vessel according to their preferences and most suitable to their purpose, or which offers the lowest freight-rate.

In accordance with the declared policy aims, the most basic effect of restraints is a general deviation from a market-orientated system of maritime transport, and a tendency towards substituting an alternative system based on differing criteria (for example, the "New International Maritime Order" as proclaimed by UNCTAD and the developing countries).

From an economic point of view, there is, therefore, a general presumption *a priori* that all those instruments and devices that are applied in order to attain the goals of restrictive policies, will have an overall effect on the economic efficiency of the shipping system, and lead towards unproductive use of resources, and to higher costs. Before discussing the various ways in which this may occur it must be said that flag-discrimination or other types of restraints do not lead to complete breakdown of sea-transport. Traditional connections, shipper preferences, or company participation may however be interrupted or seriously affected. From the numerous areas which can be affected by restraints, three have been selected, namely, vessel operating costs (both short and long-term), freight rates, and supply tonnage. Some light will also be cast on the role of regulatory policy in this context.

Effects on vessel operating costs: Apart from those discriminatory levies that are directly cost-raising, the costs of vessel operation are affected by a large variety of protectionist devices. For example, flexibility is obstructed where inter-regional arbitrage operations are prohibited by cargo reservation schemes. This will result in more ballast voyages, combined with enforced lay-up or low utilisation during off-season, etc.

The splitting-up of markets into flag-bound segments artificially limits available cargo volumes in the protected as well as the free market segments, and either lowers the utilisation of existing ships, or causes shipowners to employ small vessels with higher unit-costs. Overall utilisation is also affected where wayport traffic is prohibited, and cargo flows are segmented and bilateralised in favour of the trading partners while formerly ships used to call at ports of various countries on a given voyage. They may now have to run part-empty after leaving their first discharging port.

Where government cargo-allocating boards control cargo flows, especially under waiver systems, fluctuations may be artificially created; such bodies tend to allocate to national-flag ships more than the reserved percentage of cargo during, say, the first half of a year. Foreign ships will then have to run under-utilised while the opposite is true during the rest of the year.

Such effects mainly relate to short-term operating costs of shipping. In the longer term, there may also be a tendency of protectionist policies to interfere with technical and organisational progress. An example is Brazil; introduction of the container system was held up considerably. Also, the step into intermodalism may be hampered by government intervention even if this is due to monopoly control rather than protection of the national fleet. A tendency towards larger ships which could offer lower unit costs may be blocked if cargo reservation and market segmentation keep available cargo volumes artificially small, and insufficient to fill large vessels. In general, such policies will conserve obsolete fleet structures or operational patterns, associated with inefficiency, complacency, and a lack of progress on the side of shipping operators.

Closure of shipping markets to competition and the concomitant segmentation entail still other cost-raising disadvantages. Firstly, especially in countries which are newcomers to the maritime scene and are trying to protect their emerging fleets, there is often only one liner company which enjoys a state-protected monopoly. This will eliminate the need to rationalise, and also keep rate levels in the relevant conferences higher than necessary. Only in a few instances has this been recognised by the governments responsible for protectionist policies, and has led them to admit at least two national carriers to a trade, in order to uphold a degree of competition in the closed market sector.

Secondly, market closure will frequently be associated with elimination of third-flag operators. This is obviously the case where all cargo moving between two trading partners is completely reserved to their own fleets on the basis of 50:50 clauses. But it is only little less significant where some other allocation formula is applied, such as the 40:40:20 rule of the UN Liner Code. The economic effects are particularly unfavourable in such cases because cross traders who cannot rely on any sizeable national or home trade of their own country are often especially competitive. Empirical observation has shown that in many cases they had to cease operations though they were clearly more efficient and able to offer lower freight rates [11].

Thirdly, flag protectionism and cargo sharing in the liner trades may lead towards closure of market for all non-conference lines. This enhances the market position of conferences, and their potential for monopolistic behaviour and abuse of their dominant position. It is in obvious contrast to the traditional maritime countries' view that (closed) conferences should be subject to at least potential outside competition, and is thus of particular relevance in the context of the European competition rules. It has also been a central issue of the US-CSG shipping policy talks which have been held since mid-1982, and where closure of a (liner) market is the core.

Finally, the concept of "national carrier" which is, for example, part of the UN Liner Code, could be used to exclude lines from participation that operate as independents, or try to enter the market as newcomers. Likewise, certain lines can be protected by granting them "national carrier" status, on other grounds than ability to offer reliable services.

In all these cases, the result will be increased costs of service and a loss of efficiency which will enforce higher rates. In general terms, more ships may be required because their period capacities are artificially reduced, or vessels will be employed with higher unit costs than under competitive conditions.

Effects of restraints on shipping rates: Freight rates, apart from an influence of higher costs or from direct intervention into rate-making, can be affected in various ways. Promoting national fleets via subsidies usually leads towards downward pressure on rate levels because it tends to increase the available tonnage supply. If this occurs in a confined and isolated market sector, considerable overtonnaging with concurrent low rates may be created. Cargo sharing measures have twofold effects: they will presumably raise rates where market access is restricted to higher-cost protected fleets. But they will equally cause rates to fall on markets which are free from restraints, as relatively more ships will be offered there. Two or more distinctly different rate levels may emerge, sometimes even in the same trading area. Rate differentials may run as high as two or three times the prevailing free market rate as has been evidenced in the past by US grain exports to Egypt where part is reserved to US-flag ships under PL 480.

A particular effect on rates may occur in liner shipping if cargo-flow control in favour of national ships results in loss of high-rated cargo. This may erode revenues of foreign-flag liner services, and thus endanger their economic viability even though they will usually be more competitive than the protected fleets which profit from such measures. Similar effects may result where liner-rate regulation requires certain prolonged minimum intervals between rate increases. This may, especially under inflationary conditions, cause pressure on profitability, and probably on long-term economic viability of the lines as well as deterioration of quality of services, etc.

Effects on supply: Apart from effects on freight-rates on ship costs (short and long-term), and on competition in closed markets, there are also effects on the supply of shipping. It has frequently been pointed out that government measures to regulate markets in favour of national companies may add to excess supply as traditional maritime companies are replaced by national-flag vessels most likely through new buildings.

It is often assumed that such ships turn towards those markets that are left free from intervention, thereby increasing the available tonnage supply, and causing rates to fall (as shown above). From this it is concluded that a permanent structural crisis would develop in the free sector of the shipping markets (where the emphasis is on "permanent"). This would be very similar to the outcome of subsidies which lead to an artificial increase in the supply of ships, in excess of what is required by the demand for sea transport services. This may be the case - and is even likely to be. There are some doubts, however, as to the probable duration of such imbalance caused by excess capacity. Some kind of adjustment of shipowners to the changing prospect of profitability in the free market section can probably be expected which would cause them to refrain from ordering new ships, or to break up part of their existing tonnage in order to redress the balance, and to enable rates to rise to profitable levels again. It would seem doubtful, therefore, that the result of market closure will be a permanent crisis. Pressures on "open" shipping markets which have been observed during the last few years, and might seem to be evidence of such a relationship, are in fact mainly the result of persistent oversupply of tonnage stemming from several other causes such as shifts and slow-down of demand, subsidies to over-sized shipyards, etc.

Other effects on tonnage supply will result from interference by public authority with rate-formation. Maximum rates may lower profitability and in the longer term reduce supply. Also, enforced changes in rate-structures may lead to lack of space for lower-rated cargoes. Lower rates for LDC export cargoes at the expense of higher rates for their imports will probably cause shifts in the industrial division of labour (as the LDCs intend to bring about).

Effects of regulatory policy: There is one more point which is particularly sensitive. This is a possible connection between protection and regulatory policies which generally are meant to institute control of conferences. For example, an ambivalent role is played by the regulatory policy of the United States. According to the policy declaration of the Shipping Act of 1984, its aim is not only to control liner conference behaviour but also to promote the interest of the US merchant marine. [12] This can clearly be seen from the powers given to the FMC under Section 13 (b) 5 of the Act. The Commission may, on the pretext of some action by foreign carriers or governments which is deemed to be detrimental to the interest of either the US economy at large, or the US liner fleet in particular, unilaterally impose sanctions on such carriers or flags operating on US foreign trade routes, even if such action has taken place in trades outside US foreign commerce.

Problems have also surfaced in the context of European competition rules for liner shipping. [13] For example, shippers as well as at least one government have suggested that loyalty arrangements should be abolished or significantly limited though this may interfere with efficient, low-cost operation of liner services. Also, safeguards for outsiders have sometimes been demanded by shippers which would amount to more market segmentation and cargo reservation because such safeguards would have to ensure actual presence of at least one outsider. So far only unrestricted potential access to the liner markets for any newcomer outside the conference is held a necessary condition for allowing conferences to operate.

Another aspect of regulatory policy which may result in interference with efficient shipping organisation and higher costs is freight-rate control, such as fixing of rate ceilings, tariff-filing, setting of minimum rates in conferences to enable national carriers to compete in spite of higher costs, interference with the tariff structure on the ground of alleged "cross-subsidisation", etc. Regulatory policy in the United States has also interfered with technical progress because intermodal carriage was hampered by the FMC prohibition to include overland transport costs in a single door-to-door freight rate. Also, where new patterns of rational liner organisation emerge which affect the competitive position of seaports because they lead towards more port concentration, such changes have repeatedly been blocked.

From the foregoing it can be concluded that protectionist and promotional government action, combined with certain aspects of regulatory policies, results in numerous effects which have a direct bearing upon the efficiency of the maritime transport system, and contribute to a basic obstruction of the working of the market steering mechanism. In more general economic terms:

- they tend to impair allocative efficiency; wrong signals for investors may lead towards inefficient restructuring inside the shipping sector, or towards over-emphasis on shipping relative to other productive sectors;

- the required amount of transport services is produced with too high input of resources, and thus, at too high cost to the user of shipping services, because more ships than necessary, or the wrong ships, are used to carry the cargo, and especially because low-cost operators such as cross-traders or non-conference vessels are banned from the markets;

- hampering technical progress in shipping results in higher-than-necessary long-term costs; and

- interference with a division of labour based on comparative advantage implies a general shift towards higher-cost producers.

It seems that total available shipping space in the free markets will not necessarily be permanently larger than required because this would presume that no adjustment would be made in the longer term. On the other hand, the need for protection of national fleets may even increase because the protected carriers fail to lower their cost of operation as their competitors in the free markets are forced to do. [14] From an economic viewpoint, then, restraints on competition in shipping have definite disadvantages. They are undoubtedly impediments on the free flow of shipping services which is a contribution to rational allocation of productive resources and to an efficient division of labour. It may be recalled that the OECD countries have pledged to refrain from such impediments under the OECD Code of Liberalisation.

There are, however, a number of arguments which are put forward by those in favour of protection. Some may be mentioned:

- doubts exist as to the unambiguity of the theoretical foundation of traditional allocation theory, and accordingly to the economic justification of outright elimination of protective policies;

- income distribution is influenced to the advantage of the protecting country by participation in ocean carriage;

- national fleets are deemed essential for reasons of military or economic security;

- subsidies may accelerate technical progress (which leaves the losses associated with premature obsolescence of existing ships out of consideration);

- establishment of national fleets under protection is necessary in order to break monopolistic market dominance by liner conferences and to avoid abuse of their market power;

- a national fleet generally assists national foreign trade as well as the economy, for example, by creating backward links;

- cargo preference and subsidies must be seen as a substitute for infant-industry protection by tariffs;

- protection of shipping may compensate for protection by other countries in other sectors of the economy, e.g. agriculture.

As far as restrictions on coastal traffic are concerned these must be placed in the context of regulation of inland transport.

4 The actual impact of restrictions

So far the economic consequences of restraints have been analysed from a theoretical viewpoint. While it seems clear that restraints have unfavourable effects, the actual impact of restrictive policies on the shipping markets varies greatly. It is dependent, for example, on the degree to which market closure can be made effective - that is, can measures taken by authority be bypassed by shipowners or shippers?

Furthermore, the effect of a given measure may be dissipated if those ships that are directly affected in a certain market can "escape" to other markets. Thus, the impact of flag discrimination or of subsidies will be the greater the more their effects are confined to a specific, sharply delineated market - but it will diminish, and finally "vanish", if the various regional or goods-type shipping markets are still interdependent. For example, in the case of subsidies, the effect of additional tonnage being brought onto a market can be considerable if existing vessels cannot move to other markets. But it may be reduced to near zero if they can - because then the percentage increment of supply will be much smaller.

Effects will, accordingly, be felt most strongly if ocean shipping is organised entirely according to the bilateral

50:50 formula of cargo sharing. Obviously, therefore, the impact of market fragmentation is especially strong where trades are heavily unbalanced, so that ships are usually dependent on finding subsequent employment on other markets.

A conspicuous example of interference with commercial vessel operations is combined carriers. By combining tanker and bulk voyages in different trades which often include cross trading, utilisation of carrying capacities is to be improved. Closure of such markets may then make it impossible to load oil, and, say, iron ore on subsequent legs of one round voyage, and ships are forced to run in ballast instead, with higher costs per ton-mile performed.

The UN Liner Code is another striking example of differences between the face value and the actual extent of restrictive effects.[15]

- On the one hand, it is obviously protectionist, and it will create shifts away from an efficient organisation of liner shipping.

- On the other hand, however, the Code is applied to only a relatively small proportion of all international liner cargo. Under the so-called "Brussels Package" (EEC Regulation 954/79), the EEC Member countries have pledged not to apply its cargo-sharing provisions among each other, and also - on reciprocity - amongst themselves and other OECD member countries.[16] Furthermore, the United States has opposed the Code from its inception so that most US liner trades are free from restraints and reservation (there are some exceptions such as the US-Brazil and Argentina treaties, and the preference cargoes).[17]

Moreover, the Liner Code refers to liner conferences, not to liner trades as such. Of course, the Code itself, with its 40:40:20: formula, was a step away from strict autarchy and bilateralism under the simple 50:50 rule.

Further, and importantly, the impact of restraints may vary greatly among the different sectors of shipping. There are considerable differences between the regular and the irregular trades, especially so because liner vessels are usually employed on a round-voyage pattern. Loss of flexibility because of market closure will, therefore, have less severe consequences than in the tramp trades where no predetermined pattern of routes and employment exists, and which rely heavily on "multi-angle" traffic for efficient operation. Enforced costly ballast voyages are thus less likely in liner trades.

UNCTAD has argued that a Bulk Cargo Sharing Code would be applicable without creating all the disadvantages of cargo reservation. It has pointed to certain types of so-called semi-liner trades where, for example, bulk vessels are operating on a more or less regular "shuttle-type" basis between fixed ports of load and discharge over considerable time periods. Empirical investigation has, however, shown that considerable differences remain compared with liner operations so that any interference with free acquisition of cargo in non-liner traffic will give detrimental results.[18]

On the other hand, there are numerous mitigating factors.[19] Especially in liner shipping there is often commercial accommodation of the conditions created by restraints which serves to cushion the impact of intervention. For example, market segmentation may be overcome at least in part, by such devices as joint services of various national lines or formation of international consortia. Regional groupings may be formed with global preferences instead of strictly national or bilateral allocations.[20] One-way restrictions may be suspended under government-imposed conference statutes as in the Europe-Brazil trades. Furthermore, many countries imposing restrictions do not possess a national fleet sufficiently large so that chartered foreign-flag vessels are treated as national with regard to cargo-sharing. Many governments seem to have realised meanwhile that restrictions are the cause of disadvantages, or that the benefits from protecting a national fleet are smaller than expected so that pertinent legislation is often not strictly enforced.

An evaluation of the impact of restraints, and of the necessity to deal with them, must not overlook the influence of long-term changes. Especially in liner shipping there is a growing tension between restrictive policies and new market developments which tends to limit the applicability of restraints and thus their impact. On the one hand, traditional liner conferences are being challenged by new concepts such as the "Round-the-world" services. Such services of necessity transcend historical conference areas (though these were not separated by watertight barriers, and have not been an obstacle to outsiders). On the other hand, containerisation and other causes have brought about changes which have led the market share of liner conferences in many trades to shrink dramatically, often to just around 50 per cent.[21]

Thus, restrictive policies may loose much of their hypothetical impact unless they are extended beyond the reach of conferences. (It must, however, be noted that the effects which the decline of conference shares will have are not clear-cut as they can give rise to renewed attempts by LCDs to do exactly this). Similarly, changes in technology and organisation may reduce the impact of restraints on wayport traffic where foreign-flag vessels are no longer (or to greatly reduced extent) able to compete because of bilateral restrictions. The reason is that under the conditions prevailing under traditional break-bulk liner shipping, additional loading and discharging in wayports was profitable, but that such operations may prove too expensive in container trades so that port rotation, number of ports-of-call, etc. are often being rationalised.

Outside the liner trades, large-scale cargo reservation under an international convention similar to the Liner Code is apparently not an urgent issue at present. However, there are tendencies which may lead towards some degree of market closure and other flag-discrimination. Bulk cargoes are sometimes reserved for national-flag ships. Also, the terms of shipment may be used restrictively (export sales made under CIF terms exclusively, the ships engaged being national-flag), and this may have the same effects as outright market closure by government action.[22] A more recent trend is to conclude agreements with LDC governments; shipping companies from TMC are granted the exclusive right to carry - or at least to affreight - up to 100 per cent of a country's exports of bulk goods such as iron ore. Such contracts have been concluded between Norwegian companies, and Liberia and Zaire.[23] Obviously, these strategies can create effective monopolies, under government protection though they are on a semicommercial plane. If they continue to spread, and especially should they be extended by a Bulk Code, the impact of cargo reservation may grow considerably.

5 Conclusions and policy recommendations

In spite of those factors which help to dampen the adverse impact of restraints on the economic efficiency of shipping, there can be little doubt that at least some degree of distortion, and of interference with the market mechanism and a rational division of labour (especially with respect to cross-traders) remains. To summarise, the following are some of the major points:

- undoubtedly, restraints on the shipping markets have considerable negative effects on allocation, both inside the shipping sector and with regard to resource input relative to other sectors of the economy;

- as a consequence, sea-transport costs under a restrictive regime are higher than in its absence;

- directly affected by flag discrimination are the costs of ship operation, freight-rate levels and structures, and the supply of tonnage; in the long run, interference with technical change in shipping seems to be of particular relevance; however, subsidies as well as freight-rate control may have similar effects;

- restrictive policies will in general be less detrimental, *ceteris paribus*, in regular shipping services such as liners, as compared to irregular services such as the typical tramp operation; nevertheless, the sum total is at present greater for liner shipping because of the more extended scope of application;

- in spite of this, it seems that only up to 25 per cent of world liner trades are actually affected at present by cargo reservation measures (including the UN Liner Code), mainly by LDCs.

Taken as a whole, sea transport will not be interrupted but burdened with artificial costs of transaction; the incidence of these additional costs is uncertain.

Restrictions may thus have less impact than might be expected; but the general presumption of unfavourable effects is well-founded in spite of mitigating factors, and corroborated by factual evidence in many cases.

Interesting, though, is the apparent loss of relevance of at least some restrictive effects which is caused by long-term technological and organisational changes in shipping. These should be carefully considered for future policy. [24]

Moreover, as a source of misallocation and distortion, restraints are overshadowed, at present, by the large-scale government assistance which has been given to the shipbuilding industries since the 1973 oil crisis. To regain equilibrium in the freight markets is at least equally important as dealing with restrictive policies.

Nevertheless, the question remains whether some corrective action should not be taken to remedy the situation regarding restraints, and if so, what action would seem feasible and appropriate.

The answer to the first question would, from the point of view of economics, undoubtedly be positive. World shipping is surely in need of a more liberal framework. This is in spite of the numerous arguments put up in favour of restrictive policies by the advocates of a New International Maritime Order. It is true that most restraints applied in this context are intended to serve aims that are considered to have some higher economic justification, and often are related to income distribution rather than efficiency of production.

But it appears that most if not all of these arguments in favour of cargo-sharing and other types of essentially flag-discriminatory policies, are not really valid if judged by their actual results. This can easily be seen, for example, if the very small impact on employment in developing countries is considered. Also, the competitive position of newly-established shipping companies in developing countries has often proved increasingly unfavourable, and there have been collapses in the recent past.

The second question, then, refers to the possible ways in which to deal with restraints. A few points may be mentioned.

First, the starting-point would apparently be the traditional maritime countries which still adhere to the principles of free ocean transport, as evidenced by their signature to the OECD Code of Liberalisation. This means mainly the West European countries who should first of all undertake to solidify the free shipping area which still exists as far as their mutual traffic is concerned. Existing obstacles including cabotage reservations, other cargo preferences, and bilateral agreements should be removed. It is important that the United States be included in this group though this would probably raise problems with, for example, the Jones Act and the Cargo Preference Act.

Second, with regard to the LDCs who are at the centre of restrictive policies, several steps must be considered: (a) The traditional maritime countries should take concerted action against all attempts to extend cargo-sharing and other discrimination from liner shipping to other sectors, for example, under the heading of a Bulk Cargo Sharing Code as envisaged by UNCTAD. There would also have to be steps to obviate unilateral action to the same effect. (b) *Also,* the TMC would have to resist any attempts by LDCs to extend restrictions which refer to Liner Conferences under the UN Liner Code - especially the cargo sharing provisions - from the conference share of a liner trade to all cargo moving in that trade. The freedom of outsiders must be preserved since it is an important element of conference philosophy of the traditional maritime countries. This is a relevant issue for the forthcoming (1988) Review Conference on the UN Liner Code. Also, the United States and the CSG countries have tried to draw up a procedure of co-ordinated resistance against restraints introduced by foreign governments (Statement by Copenhagen meeting, 30 April 1986).

These are short-term steps; their aim is to stabilise the *status quo* of LDC trades by checking the tendency of restraints to spread further. In the longer term, existing barriers should also be removed in the LDC trades. The general objective would be to rebuild a market-orientated liberal order, in principle if not in every detail. This would obviously involve a great deal of effort to persuade developing countries of the multiple advantages and general superiority of the traditional over the new maritime order.

Of course, in trying this, the traditional maritime countries must recognise the fact that the developing countries are sovereign nations, and the latter not only have a right to follow their own philosophies but they can also enforce such policies, or obstruct attempts to restore the old order one-sidedly. A measure of co-operation and compromise would, therefore, seem a precondition for success, and would correspond to the principle of comity in international law.

However, there also seems to be some reason for optimism as there is a growing recognition among LDC governments of the economic problems involved in establishing and expanding own merchant marines. [25] Perhaps it is now easier than before to persuade the advocates of a NIMO that it would be of little use to anyone in the world if the traditional, efficiency-oriented system of sea-transport were replaced by an alternative which is often in definite contrast to economic considerations.

It is nevertheless true that, to reduce to a unified system the present quadrangle of international shipping policy may appear as the squaring of the circle. But, judging by its probable merits, the attempt would be worthwhile - to provide ocean shipping with the liberal framework again from which it has benefited so much in the past, and together with shipping the world economy of which it is the foundation.

NOTES AND REFERENCES

1 For a more detailed presentation see Hans Böhme, *Restraints on Competition in World Shipping*, London: Trade Policy Research Centre, 1978 (Thames Essay No. 15), p 23 *et seq*.

2 Under Public Law 83-664 of 1954. On US shipping policy see, e.g. S A Lawrence, *United States Merchant Shipping: Policies and Politics*, Washington, D.C.: The Brookings Institution, 1966; E G Frankel, *Regulation and Policies of American Shipping*, Boston, Mass.: Auburn House Publishing Co, 1982.

3 On the development of UNCTAD's activities in the maritime field see, for example, the Secretariat's Annual Reports, and also, the Committee on Shipping's session reports. Further B Gosovic, *UNCTAD and Compromise. The Third World's Quest for an Equitable World Economic Order Through the United Nations*, Leidon: A W Sijthoff, 1972, p 3 *et seq*, and 138-52 - L Juda,

World Shipping, UNCTAD, and the New International Economic Order, *International Organization*, Madison, Wisc., Vol 35 (1981), No 3, p 493 *et seq*. One of the leading countries to implement a comprehensive maritime policy based upon these considerations was Brazil since the late sixties.

4 This refers mainly to the European Community countries - viz. the discussion of the "competition articles" of the Treaty of Rome and liner shipping. Under Community competition rules liner conferences shall be subject to more public control and regulation than traditionally. (See European Commission, *Proposal for a Council Regulation Laying Down Detailed Rules for the Application of Articles 85 and 86 of the EEC Treaty to Maritime Transport*, com(81) 423 final, Brussels: Commission of the European Communities, 1981).

5 See, European Commission, *Progress Towards a Common Transport Policy: Maritime Transport*, Commission communication and proposals to the Council transmitted on 19 March 1985 (based on COM(85)90 final), Bulletin of the European Communities, Brussels, 1985, Supplement 5/85, p 7 *et seq*.

6 These talks developed from the long-standing dialogue between the United States and the Consultative Shipping Group countries mainly on questions of US liner-shipping policy. In June 1982 the US raised problems of market access created, in its view, by the acceptance of the UN Liner Code. A short common statement on principles of future policy to be followed by the US and the EC was issued at the meeting at Copenhagen on 30 April 1986, to the effect that market access shall be kept open, and attempts by developing countries to close liner markets shall be resisted jointly by the two groups.

7 For example, a trend towards long-term transport contracts has been discerned in liner business - especially after the US Shipping Act of 1984 substituted service contracts for loyalty ties of the dual-rate contract type - which may eventually replace conference arrangements. See, Without Conferences - what? *Shipping News International* , Oslo, 1985, No 5, p 54 *et seq*.

8 See, for example, the author's Die Schiffahrtspolitik der Entwicklungsländer, *Die Weltwirtschaft* , Tübingen, 1971, No 1, p 177 *et seq*, and his Freiheit der Meere oder Bilateralismus in der Seeschiffahrt. Die Schiffahrtspolitik Brasiliens - Modell für

die Zukunft?, *ibid.*, 1972, No 1, p 202-31, especially p. 210 *et seq*.

9 Flag discrimination began, on a significant scale, already in the early post-war years when both the United States and certain developing countries such as Chile or Colombia, began to reserve part of their foreign-trade cargoes to their national-flag ships. See, for example, studies published as early as the mid-fifties, by the International Chamber of Commerce, and the Maritime Transport Committee of the OECD. See also O. Henell, *Flag Discrimination, Purposes, Motives, and Economic Consequences*, Helsingfors: Söderström & Co, 1956.

10 Presumably this is often overlooked when the potential effects of cargo reservation are assessed, as "50 per cent" may suggest "only a part". This fact compares with the granting of inland treatment to foreign-flag ships (that is more than customary most-favoured-nation treatment) under the traditional *Freedom of the Seas* which is still the essence of the OECD Code of Liberalisation in its application to the maritime field.

11 A conspicuous example is the reorganisation of Brazilian liner trades after 1968 where several European cross-trading liner companies had to terminate their old-established services between Brazil and the North American East Coast.

12 Shipping Act of 1984 (PL 98-237) of 20 March 1984, 98 STAT 67, Section 2, Declaration of Policy, No. 3.

13 See in detail Hans Böhme, Current Issues and Progress in European Shipping Policy, *The World Economy* , London, Vol 6 (1983), No 3, p 325-52. Also the detailed hearing held in Britain (House of Lords, Select Committee on the European Communities, Session 1983-84, 3rd Report, *Competition Policy: Shipping, London: HMSO 1983*).

14 This argument has been proposed in recent years even by carriers in the United States where subsidies have been handed out to most liner companies for several decades, and where it has now been claimed that without subsidies the relevant carriers would be sufficiently competitive to stay in business. See for a discussion of related points, Clinton H Whitehurst, *The US Merchant Marine. In Search of an Enduring Maritime Policy*, Annapolis, Md.: Naval Institute Press, 1983, p 83 *et seq*.

15 The UN Liner Code became effective on 4 October 1983. For its full text see, United Nations Conference of Plenipotentiaries on the Code of Conduct for Liner Conferences, held at Geneva from 12 November to 15 December 1973 (first part) and from 11 March to 6 April 1974 (second part), Volume II, Final Act (including the Convention and resolutions) and tonnage requirements, New York: United Nations, 1975 (TD/CODE/13/Add. 1).

16 EEC Regulation 954/79 ("Official Journal of the European Communities", Brussels, No L 121 of 17 May 1979) was devised so as to enable EEC member countries to accept the UN Liner Code without contravening the requirements of non-discrimination contained in the Treaty of Rome. For details see Hans Böhme, Current Issues and Progress in European Shipping Policy, *op. cit.* p 330 *et seq.* - There is, however, disagreement on the point whether in the liner trades between EEC countries and developing countries (where the latter have ratified the Liner Code or apply protectionist policies unilaterally) market access will in the long run be more restricted as the Code rules or some equivalent are extended from the liner conferences to the whole of the respective trades.

17 It should be noted that the Code is intended to cover *all* cargo moved in a Conference's trade, with the sole exception of military cargo destined for the own defence of the respective country. No exception shall be made for additional reservation of "government impelled or sponsored" cargo which has frequently and to a very considerable extent been the case so far in numerous countries, developing as well as developed (Article 2, paragraph 17).

18 See studies carried out by Fearnley & Eger of Oslo in 1978 (quoted in OECD, *Maritime Transport 1978*, Paris: OECD, 1979, p 74 *et seq.)*

19 The extent to which market access is actually restricted by developing countries is apparently smaller so far than might be expected. An investigation carried out by the OECD has shown that in 1983, of a total of 88 countries, only nine completely banned *outsiders*, six others having *some* restraints. Seven and three, respectively, gave a definitely or with reservations positive answer to the question whether they had any restriction on *cross-trader* entry and participation in Liner Conferences. See OECD, *List of Legislative and Discriminatory Measures Affecting Shipping*, Paris, OECD, 1983 (Document TS-5 (a)/1-2).

20 Such efforts have especially been made in West African countries where the disadvantages created by strict enforcement of the Liner Code cargo-sharing principle ("40:40:20") are particularly visible, because of the relatively small cargo volumes to and from most of these numerous countries. These volumes are mostly insufficient to sustain individual liner companies; ship sizes or sailing frequencies could be unfavourably affected.

21 See, for example, the data on Liner Trades between Developed and Developing Countries in: Cargo and Carriage, prepared for a study by Elliot Schrier and Hans Böhme for the Trade Policy Research Centre, London, 1985. For example, in 1983 only a little over one-third of the available capacity offered in the OECD developing country liner trades was provided by conference carriers.

22 This is in distinct contrast to normal commercial practice where of course some carriers (in fact, all but one) are excluded under the terms of a particular contract. This is of necessity so, but without intention whereas in the above-mentioned context the terms of shipment are used to apply flag-discrimination.

23 These are clear cases of exploitation by traditional maritime countries' shipping companies of restrictive policies or inclinations found in developing countries. There may thus be a collusion of interest which is not in the interest of low-cost, efficient shipping services.

24 This refers, in particular, to efforts made by the United States to subject liner shipping worldwide to detailed regulation in favour of "more competitive market access"; if contestability of liner markets is taken as the relevant criterion, it appears that it is greatly enhanced by current changes in liner economics and technology, and that there may be less rather than more need for such regulation (though this does not refer to dealing with protectionist government actions).

25 For example, Brazil has recently begun to readmit third-flag carriers to its liner trades from which they had been excluded for more than 15 years. See, Cost of Flag Discrimination too High, *Norwegian Shipping News* , Oslo, Vol. 40 (1984), No. 11, pp 32 and 56.

CHAPTER 3

THE EEC COMMON MARITIME TRANSPORT POLICY AND THE LIBERALISATION OF WORLD SHIPPING MARKETS

John Tzoannos

1 Introduction

The purpose of this chapter is to identify and assess the implications of the specific policy measures on maritime transport proposed by the Commission and partially endorsed by the Council [1] in relation to the conflict between liberalism and interventionism in international maritime affairs.

It is worth noting in this respect that following a long period of continuous erosion of the conditions of free competition in the international shipping markets there is now tangible evidence of a reaction to protectionism and flag discrimination practices from the Western developed countries, through coordinated initiatives aiming at the creation of more liberal conditions.

Another important development is that the maritime sector appears now to enjoy an upgrading in government thinking in many countries in terms of its importance. This may be due to the effects of the prolonged shipping crisis which has highlighted the significance of shipping as an earner of foreign exchange and generator of employment. Shipping is looked upon as a sector much more on its own merits rather than as a trunk industry subservient to other sectors.

The chapter opens with a brief review of the developments concerning the conflict between liberalism and interventionism in shipping. It then examines the relevant action by the EEC prior to the production of the 1985 proposals. This is followed by a presentation of the key elements in these proposals which might have a bearing on the process of liberalisation. Finally, there is an evaluation of these elements in relation to the developments described in the subsequent sections.

2 Liberalism versus intervention

In order to assess the international impact of the EEC maritime policy it is necessary firstly to review the state of affairs concerning the conflict between liberalism and interventionism in the international maritime transport markets.

This distinction between liberalism and interventionism has a broader meaning than the dichotomy between free trade and protectionism which is more commonly used. [2]

Interventionism apart from the various cargo reservation or sharing practices includes all forms of subsidies given by governments to ship operators; the tying of economic aid granted to various countries with the use of the national flag of the donor, and the support of commercially derived impediments to free competition, such as the system of closed conferences.

An extensive inventory of the practices followed by governments in various parts of the world which create difficulties in relation to free competitive access in maritime transport has been produced by the OECD. [3] This inventory does not however include commercially derived impediments to free competition in shipping. The existing acts of interventionism can be classified into broad categories as follows:

(a) Unilateral cargo reservation and flag preference

These acts include such measures as government controls concerning the nationality of the vessels that are entitled to carry cargoes in the external or domestic trades of the country, preferential shipping clauses in trade agreements, port regulations or dues which discriminate in favour of the national flag vessels, etc.

Such acts are to be found mostly among developing countries than developed ones, although quite a few of the latter, including members of the EEC, are not far from blame.

(b) Bilateral division and allocation of cargo

This form of interventionism is usually institutionalised in the context of bilateral trade or economic cooperation agreements between two countries, whereby the exporting and the importing country reserve for their own fleets the cargoes generated in their trade.

Such bilateral agreements are common among developing countries usually of the same region, but also between certain OECD and EEC member countries and developing countries.

Bilateral allocation of cargoes is common also in the direct trade with Eastern Bloc countries on the basis of shipping agreements which have been concluded with them by many developing as well as OECD countries.

Under this heading one should also include the bilateral cargo sharing system institutionalised in the liner conference trades with the Code of Conduct for Liner Conferences which came into force in 1983 and is a result of pressures from developing countries. That system constitutes an allocation of cargoes in the liner trades between two countries on a 40% - 40% basis for the importing and the exporting country respectively, while 20% is left open to cross-traders.

(c) Activities of state trading countries

The usual practice through which these countries, primarily the USSR, allocate cargoes to their national flag vessels is by selling their products on a CIF and buying on a FOB basis.

This practice is reinforced with a network of freight bureaux established in other countries which direct cargoes to the state trading country's own fleet.

An additional impediment imposed on foreign shipping companies is that these are prohibited from setting up their own agencies on the state trading countries' soil, and thus depend on the latter for cargo accessibility.

(d) Government subsidies to shipping

These include any payment by the public sector of an economy which affect directly relative prices either in the services or the factor markets of the national shipping sector. [4]

Subsidies can be classified into three main categories: (i) tax concessions, (ii) direct cash transfers, and (iii) provision of cheap credit. [5]

These acts of government interventionism are widespread throughout the world. An extensive inventory has also been produced by OECD. [6]

(e) Toleration or support of commercially derived impediments

This category concerns restrictive practices by commercial enterprises which tend to eliminate competition in the shipping markets, mainly through the participation in closed conference.

Support of closed conferences is to be found among Western European maritime countries, while the US and Greece are opposed to them.

It should also be pointed out that the restriction of access to cargoes as a result of closed conferences had been one of the main arguments used by developing countries in pressing for a Code of Conduct through the UN.

The main sources of pressure that counteract the trend towards interventionism in shipping are as follows:

(a) The OECD Code of Liberalisation of Current Invisible Operations adopted in 1961, and in particular note 1 of its Annex A.

This constitutes an undertaking among OECD member states to eliminate between one another restrictions that hinder the provision of maritime transport services by the residents of one member state in another.

This undertaking is a manifestation of the inclination of developed western countries in favour of liberalism in trade, but it must be pointed out that it has not eradicated restrictive practices among them, since it does not provide for enforcement procedures.

In addition the application of the OECD Code is limited to the relations between developed Western countries and does not cover their relations with the third world or the Comecon countries.

(b) The coordinated resistance shown by the developed, market-economy countries, known as the "B-Group" in the UNCTAD to the attempts by the "Group of 77" (made up by developing countries) to extend the institutionalisation of cargo sharing in the field of dry and liquid bulk markets.

(c) The CSG-US [7] dialogue, which is starting again, and aims at coordinated resistance against other countries that practice cargo sharing and flag discrimination. This dialogue has not so far produced any concrete results in terms of an agreement for joint action by the developed Western countries.

This may be attributed to the fact that these countries are themselves not unequivocal in the practice of liberalism, but rather their actual policies represent a blend of liberalism and protectionism. This has led to a series of acrimonies which have slowed down the process of reaching an agreement.

For example a major source of friction has been the bilateral cargo sharing agreements that countries participating in the above dialogue (including the US) have concluded with third countries. The US is unhappy about certain European countries having ratified the UN Code of

Conduct, while members of CSG are critical of the bilateral agreements between the US and certain Latin American countries or the practice of cabotage. [8]

Nevertheless, the existence of that dialogue has produced some indirect results in favour of liberalism in the sense that it has put pressure on the Western countries to refrain themselves from further acts of protectionism in shipping and to liberalise their own markets for shipping services.

Furthermore it has encouraged an ad hoc coordinated reaction to new cargo reservation practices by third countries. A good example is the successful resistance to such a move by Indonesia.

(d) The US maritime policy. The US being the single most important world economic power exerts a significant influence on the international maritime scene.

Although their fundamental economic philosophy is based on the principles of liberalism, the US maritime policies have developed since the beginning of this century in a haphazard way, so that today they consist of a mixture of free trade and interventionism. [9]

On the one hand the US reserves under the Jones Act (Merchant Marine Act, 1920) coastal shipping for US owned and built ships, has concluded cargo sharing agreements with Latin American countries and is negotiating a new one with the USSR [10] but on the other it prohibits closed conferences under its antitrust laws or it opposes the UN Code of Conduct.

Furthermore, the US operates a system of subsidies for the construction of vessels in the national shipyards and for their operation under the national flag.

However, since the Reagan administration took office in 1981 there has been a tendency to diminish the role of subsidies and to decrease the degree of government regulation of shipping. The latter tendency has found a formal expression in the 1984 Shipping Act which simplified American regulations on shipping conferences which had in the past become extremely burdensome.

That Act allows both US and foreign line operators to join *open* conferences without fear of prosecution under the anti-trust laws, but stipulates that conferences must allow any regular liner operators to join or leave them at will.

Another interesting aspect of the Act is that it has cut back the extra-territorial powers over liner activities, given to the US administration by previous legislation, and which had created serious friction with the governments of other Western countries.

Finally, there have been signs in the recent past of a strengthening of the American resistance to discriminatory practices by third countries involving the use of the threat to ban vessels of the offending country from the US trades.[11]

(e) The US initiative to bring trade in services, including maritime transport, under GATT. This initiative has met the strong resistance of the developing countries, headed by India and Brazil.

In that respect it is worth noting the existence of the 1984 US Trade and Tariff Act, which is meant to increase the American negotiating power in international fora to force negotiations in several areas of trade, including services, by enabling the US administration to penalise tradewise those that refuse to negotiate.

3 The EEC initiatives

EEC initiatives in the international shipping scene have been a phenomenon of the last decade only, and they represent attempts at solving individual problems instead of an overall arrangement.[12] The legal instruments issued concern the coordination of the maritime policies of member states in the international fora, the monitoring of the activities of the liner shipping companies of the Eastern bloc countries in certain trades, measures to improve maritime safety and the prevention of pollution, and the ratification of the UN Code of Conduct.

In terms of the liberalism vs. interventionism dichotomy the ratification of the UN Code in the context of an

EEC regulation is by far the most important instrument issued so far.

When the UNCTAD IV session took place in 1974 there had been no EEC representation in the deliberations, nor indeed a common stance by the then member states. Just before the final act, the scene changed by the well known judgement of the European Court of Justice on the *French Seamen's Case* - which altered drastically the thinking on the applicability of the general provisions of the Treaty of Rome to shipping. It became clear then that for member states to sign the Code individually would contravene the provisions of Articles 113 and 116, which refer to the common commercial policy.

Three member states (Belgium, France and Germany), did in the end sign the Code, despite the Commission protestations to the contrary.

Following protracted negotiations a compromise between the different views of member states and the Commission was found in Regulation 954/1979, known as the *Brussels Package.* [13]

The regulation aims at making the accession to the UN Code by a member state compatible with the general provisions of the Treaty of Rome, such as the right of establishment, the competition rules, non-discrimination, and the common commercial policy.

Basically, it obliges member states to ratify the Code subject to the following reservations:

(i) The Code in its entirety will apply to the liner trades between developing and developed countries.

(ii) The 40-40-20 formula envisaged by the Code will not apply in the trades between EEC countries, and on a reciprocal basis between EEC and other OECD countries, and,

(iii) The share allocated under the Code to the EEC lines will be redistributed among them on the basis of commercial criteria.

Another important aspect of Regulation 954/1979 is that it accepts closed conferences, through the provisions of its Article 2.

On the other hand, the Commission and the Council have tried with successive proposals since 1981 [14] to regulate the behaviour of liner conferences by laying down detailed rules for their exemption from the application of competition rules enshrined in the Treaty of Rome (Articles 85 and 86).

The proposals contain detailed rules of procedure which grant extensive powers to the Commission for investigations into the behaviour of conferences and for imposing sanctions in the event of an infringement. Also, they introduce the possibility for the shipper to be able to choose between immediate or deferred rebates in the commitment of cargo.

Subsequent drafts have also introduced the condition that no loyalty agreement may tie more than 70% of the value of cargo generated by shipper so that he may enjoy rate rebates. [15]

Nevertheless these draft proposals continue to be based on the view that the system of liner conferences offers organisational advantages to seaborne trade in terms of regularity of service and stability of rates. [16]

Also of interest are the Council Decisions on the collection of information concerning the activities of carriers participating in cargo liner traffic in certain trades (19/4/EEC, 81/189/EEC, 82/870/EEC, and 84/656/EEC).

These have aimed at identifying the extent of the damage caused to EEC shipowners by the activities of eastern bloc fleets and at warning signalling to the governments concerned that the EEC is willing to react to what it considers to be unfair practices on their part.

In addition to the above instruments, which have been produced with direct reference to shipping, there exist certain trade and economic cooperation agreements among the numerous ones that have been concluded between the EEC and third countries which have a bearing on shipping.

These agreements are the Lomé conventions and the economic cooperation agreements with Brazil (1980), the Andean Pact countries (1983) and China (1985).

The Lomé II Convention (1979) contained a joint declaration between the EEC and the participating developing countries whereby "the Community recognises the aspirations of ACP States to have a greater share in the carriage of bulk cargoes".

This declaration, which indirectly accepted cargo sharing, was a result of pressure exerted by the ACP countries and the wish of the EEC to satisfy their aspirations in the context of the North-South dialogue.

This acceptance of interventionism in the maritime relations between the EEC and the ACP countries, has been modified considerably in the Lomé III Convention (1984). The relevant title there (Title V), involves an acceptance by the parties concerned of a more liberal regime in maritime transport.

For example, Article 86 of the Convention contains a declaration that the objective of cooperation between the EEC and the ACP countries in the field of shipping "shall be to ensure harmonious development of efficient and reliable shipping services on economically satisfactory terms by facilitating the active participation of all parties according to the principle of unrestricted access to the trade on a commercial basis".

There is still in the new Convention a statement acknowledging the aspirations of the ACP states for greater participation in shipping and in particular bulk cargo shipping but this is linked to the ideas of efficiency in the shipment of cargoes and of the preservation of competitive access. (Article 88)

This overall shift in proclaimed intentions may be attributed to a hardening of EEC attitudes concerning protectionism in maritime transport by third countries as a result of pressure by the governments of member states (which strongly opposed cargo sharing in the bulk trades under UNCTAD) as well as by interest pressure groups (CAACE, UNICE).

The reference to shipping in the context of the trade and economic cooperation agreements with Brazil and the Andean Pact countries, does not go beyond an exchange of letters of intent that the problems arising in the maritime relations between the EEC and the countries concerned will be examined in a cooperative spirit. Therefore, this reference is of little practical use.

The cooperation agreement with China constitutes a general framework within which special agreements on specific economic sectors, such as shipping, are to be concluded. The EEC Commission is in the process of preparing a draft text for such an agreement in consultation with interested economic groups within the Community.

It can be seen from the above that the acts by the EEC so far in the sphere of maritime transport represent a blend of both liberalism and interventionism.

For example the Brussels Package accepts the principle and practice of cargo sharing, but it attempts on the other hand to limit that practice only to the trades between the developing countries and the OECD members. Through the principle of reciprocity it attempts to preserve a liberal regime in the trades between the EEC and the other OECD member states.

However, that latter aim is undermined by the acceptance of the restrictive practices of closed conferences, or the existence of government subsidies to shipping companies within the EEC.

This equivocal approach is to be seen also in the shipping clauses contained in the Lomé Conventions.

This approach might be attributed to (a) a desire to preserve the unity of the EEC camp which came under considerable pressure in the case of the accession to the Code and (b) a fear that if no compromise is found with the developing countries these might proceed on a unilateral basis to even more extensive protectionist practices.

It should also be noted that among the plethora of trade or economic cooperation agreements concluded by the EEC and third countries the ones with shipping clauses are very few and without any 'real teeth' in preserving a competitive environment in the relevant trades.

Finally, it should be pointed out that instruments already produced by the Community have introduced the idea of reciprocity in extending the same treatment received by Community shipping to that of third countries (e.g. the rest of the OECD).

Thus, the prevailing thinking is that there should be no 'unilateral disarmament' in terms of government imposed impediments to competitive access in maritime transport.

It is against such a background that the Commission proposals for a Common Maritime Policy were produced in 1985.

4 The Commission memorandum and the Council decision

The key element in the Commission memorandum, as seen from the angle of the present analysis is that it adheres to the view that a maritime policy based on a free trade approach is in the best interest of Community shipping.

The decline in the percentage of the world tonnage represented by the Community fleet is attributed to a loss of comparative advantage (due to internal pressures) as well as to the growth of protectionist or unfair practices by third countries.

In relation to the position adopted *vis-à-vis* the discriminatory practices of third countries, the Commission stresses the need for liberalising the internal EEC market in shipping services, by applying the principle of the freedom to provide services (Articles 59, 60 and 61 of the Treaty of Rome) to maritime transport.

The Commission also continues to subscribe to the view that the liner conference system is beneficial to world seaborne trade, in terms of providing stability of freight rates and regularity of service, and therefore believes that there should be legal provision for its exclusion from the application of the competition rules of the Treaty of Rome (Articles 85 and 86).

The Commission's communication contained, in the form of an annex, six concrete proposals:

The first proposal (Annex II-1) specified the required Regulation coordinated action by member states against third countries in the liner or bulk trades so as to safeguard free access to cargoes in the ocean trades by shipping companies of the EEC or the OECD.

Coordinated action is to proceed in two stages: (a) diplomatic measures, and if these fail, (b) counter-measures directed at the shipping companies of the offending country. The counter-measures may consist of a partial or total restriction of access to cargoes or of financial charges. There is immunity, however, from the EEC counter-measures if the third country restricts access under the provisions of the UN Code.

The second proposal (Annex II-2) referred to the application of the principle of freedom to provide services to sea transport. The application of the principle involves the addition of restrictions to the provision of such services by persons (legal or physical) established in a member state in (a) the trades between any two member states, (b) the coastal trades of a member state (cabotage) including its overseas territories, (c) the trades between a member state and third countries (cross-trading) and (d) the supply of off-shore installations.

In the case of cabotage the abolition was envisaged to take place by 1992, i.e. in line with other measures establishing the internal market. During these transition periods no member state was to be permitted to make existing provisions restricting access to cargoes less favourable to persons established in other member states.

The third proposal (Annex II-3) concerned an amendment to the existing Decision 77/587/EEC setting up a consultation procedure among member countries on their maritime relations with third countries or on their action relating to maritime affairs in international organisations.

The aim here is to make the consultation procedure more meaningful by obliging a member state that is negotiating shipping agreements with third countries to consult other member states prior to the conclusion of the agreements so that these agreements respect the measures adopted jointly in pursuance of Community maritime policy. Indirectly this decision also aims at discouraging the conclusion of such agreements by individual member states.

The fourth proposal (Annex II-4) clarified the definition of "national shipping line" for the purposes of the UN Code of Conduct.

The aim of the directive is to prevent the abuse of the criterion of the national shipping line by companies belonging to nationals of third countries by imposing additional criteria to those imposed by the Regulation 954/79.

The fifth proposal (Annex II-5) contained amendments to the 1981 draft Regulation laying down rules for the application of the competition rules to liner trades. Tramp vessel services are excluded from the scope of the Regulation.

These amendments make the conditions for the exemption of liner conferences from the competition rules more lenient. For example according to this proposal the abolition of the provision that the maximum commitment of cargo by a shipper under a loyalty agreement should be 70% which was to be found in the earlier drafts.

Furthermore, the notice period for the termination of loyalty agreements was not reduced from 6 to 3 months as demanded by shippers organisations.

The final proposal (Annex II-6) dealt with unfair pricing practices in cargo liner shipping practised by shipowners established in a third country.

'Unfair pricing practices' are taken to mean the continuous charging of a freight rate on a particular shipping route which is lower than the lowest rate charged during at least one year on the same route and for the same commodity by an established and 'representative' outsider.

The draft Regulation envisaged detailed procedures for the lodging of complaints, the examination of the injury on Community shipowners, and the imposition of redressive duties on the offenders.

Following negotiations, the Council of Ministers adopted in their meeting on 22 December 1986 four Regulations as follows:

Regulation 4055/86 which corresponds to Annex II-2
Regulation 4056/86 which corresponds to Annex II-5
Regulation 4057/86 which corresponds to Annex II-6
Regulation 4058/86 which corresponds to Annex II-1

Annexes II-3 and II-4 were given a low priority in the negotiations as not representing issues of such an importance that would justify any further delays in Community action. Thus in about one and a half years the key parts of the Commission's proposals have been transformed into Community Law.

The main changes to be found in the above Regulations compared to the corresponding Annexes concern the issue of cabotage and the application of the competition rules. Cabotage is not after all included in the restrictions to shipping activity that will be abolished under Regulation 4055/86. The abolition of cabotage remained a thorny issue throughout the negotiations in the Council. At the end of the day it was decided, in a spirit of compromise between all parties concerned, to exclude it, so that the rest of the package of measures could be adopted as a matter of urgency in view of the crisis facing Community shipping.

The Regulation adopted concerning the application of competition rules to the liner trades states that loyalty agreements may be offered to shippers by conferences but not imposed. This modification to the original text is however of little practical use.

On the other hand, the same Regulation makes more explicit reference to Article 86 of the Treaty of Rome concerning the abuse of dominant position. This has resulted in a more balanced treatment of the conflicting interests to be found in the liner trades (conferences independent operators and shippers) compared to the original draft.

It should also be pointed out that the European Council in adopting the above Regulations made a statement emphasising that these mark only a first stage in the development of a Community shipping policy with the aim of encouraging an efficient and competitive Community shipping industry. In that statement the Council also calls for efforts to reduce the disparities in operating conditions and costs between the Community fleets as a whole and their foreign competitors.

5 Assessment of the new regulations

The key point emerging from the Regulations introducing the Common Maritime Policy is that these continue on the same path of blending a general liberal orientation with specific acts of protectionism, which has been followed by the Community in its past actions in the maritime sector (and for that matter in other sectors, notably steel).

Proposed actions that are clearly bound to create a more liberal regime in maritime transport are those contained in Regulation 4055/86.

Also, the original proposals contained in Annex II-3 (which was not adopted), can be considered to be beneficial to the liberal cause in the sense that they will tend to discourage the conclusion of bilateral cargo sharing agreements by individual member countries with third countries.

A regulation for the application of the competition rules to liner conferences was very much needed, especially after the UN Code and Regulation 954/79 so as to curtail their potential abuse of dominant power. However, Regulation 4056/86 offers less protection to shippers compared to the ones that have been formulated in the past, and does not create the necessary conditions for the unimpeded operation of outsiders.

The effects of the instruments introduced by Regulation 4058/86 are the most difficult to assess, since these are conditional on the reactions of the countries toward which the counter-measures are addressed.

The existence of the possibility of coordinated joint resistance to protectionism might act as a deterrent for such acts by third countries. In fact the creation of the legal framework for coordinated resistance by the CSG countries has been one of the basic American demands in the CSG-US dialogue mentioned above. The realisation of coordinated resistance in the form of diplomatic action will also act as a deterrent. But there is a possibility that the adoption of counter-measures might lead to a trade war in maritime services which would have the opposite effect to the one hoped for by Regulation 4056/86.

In fact the combined effects of counter-measures under this Regulation and protectionism under Regulation 4056/86 concerning liner conferences might lead to increased tensions with the developing countries. One of the reasons put forward by the latter in pressing for cargo sharing under a UN Code had been the impediments to access created by closed conferences. Thus, they might argue that their acts are only different forms of protectionism to those accepted within the EEC and that they are therefore unduly penalised.

It is possible, therefore, that the equivocal approach in the EEC policy might hinder the efforts that the Commission intends to make so as to improve cooperation in shipping matters with the ASEAN and the Andean Pact countries, and China. Also, it is likely that this approach will limit the credibility of the Community if it decided to bring shipping under the auspices of GATT. But, on the other hand, it may provide the EEC with a negotiating leverage.

Despite the inconsistencies of the EEC policy which might lead to tensions with developing countries the element of deterrent in Regulation 4058/86 would be enforced, however, if the threat of counter-measures was extended to the offending country's exports of goods or services to the EEC. The existence of this danger for their exports would be expected to make potential offenders more careful in resorting to protectionist practices in the first instance, or in reacting to counter-measures taken by the EEC and would thus minimise the risk of a trade war in maritime services.

An important characteristic of the Regulation is that it provides for counter-measures by the EEC when discriminatory acts by third countries affect companies in other OECD countries as well. A corollary to this provision is that no counter-measures are possible if the offending country belongs to the OECD.

Thus, the Regulation reinforces the element of solidarity in maritime affairs among developed (OECD) countries, found in Regulation 954/79. In turn it emphasises the distinction between North and South. This might further increase tensions with the developing world. This element of solidarity was also present in Annex II-4 of the original Commission proposals (which was not subsequently adopted).

The effectiveness of the Regulation is reduced by the exclusion from its scope of the restrictions to access imposed by third countries under the UN Code. This exclusion opens the door to abuses by these countries without any fear of counter-action by the EEC.

The emphasis given to the creation of a free trade area in maritime services within OECD is consistent with American thinking with respect to the ongoing efforts to bring services under the auspices of GATT. The US administration has indicated that, if these efforts fail, attention will be focused in creating a 'regional' rather than a 'global' free market among countries that practice liberalism. [17]

While the adoption of coordinated resistance measures will encourage EEC relations with the American side in the context of the CSG - US dialogue, these relations are not going to benefit from the provisions of Regulation 4856/86 since they leave intact the system of closed conferences.

It is likely also that the abolition of bilateral agreements between EEC member states and third countries under Regulation 4055/86 would put pressure on the US administration to abolish its own bilateral agreements or to refrain from the conclusion of new ones.

A potential source of additional frictions with third countries could be created by the application of the provisions of Regulation 4057/86 which can easily be turned into a protectionist instrument for Community liner shipping companies.

Ostensibly the proposed Regulation aims at fighting unfair pricing practices by companies from third countries with the argument that these practices are sustained by various forms of government protection including subsidies.

Primarily the companies in mind belong to the state trading countries or operate from Hong Kong and Taiwan.

However, the fact that an operator offers lower rates might not necessarily be due to unfair pricing as a result of government support. It could also be due to the employment of factors of production from different markets which present lower opportunity costs [18] or to a more efficient organisation in the production of transport services.

In fact in the case of liner shipping services, which are produced as part of an integrated multi-modal system, lower rates might be due to a higher degree of efficiency in the organisation of the whole system or in the operation of the land based part of the system. It should also be borne in mind that subsidies to shipping exist in many EEC countries, a fact that undermines the arguments in favour of this Regulation. How third countries are going to react if Regulation 4057/86 turns into an instrument of protectionism is difficult to say.

It is noteworthy in this respect that the Soviet Union has already reacted to plans by five EEC governments (Belgium, France, Germany, the Netherlands, and the UK) to monitor Soviet shipping in certain liner trades from March 1986 by threats of retaliation. [19]

Finally, in relation to the Soviet Union and other state trading countries it must be pointed out that the new Regulations do not contain any provisions for tackling the problem of flag discrimination through the operation of freight bureaux.

6 Conclusions

From the above analysis it may be concluded that in general the Community policy is likely to contribute to the liberalisation process within the OECD world. But on the other hand it will strain relations in shipping matters between the EEC and developing as well as the Comecon countries. This strain if it deteriorates into a war in maritime transport services is expected to create an even more protectionist climate outside the OECD world.

NOTES AND REFERENCES

1 Commission of the E.C. *Progress Towards a Common Transport Policy: Maritime Policy*, COM(85) 90 final.

2 Henderson, P D, Trade Policies: Trends, Issues and Influences. *Midland Bank Review*, Winter 1983.

3 OECD *Maritime Transport 1983*, Paris 1984, Chapter 5. Also see US Maritime Administration, *Maritime Subsidies*, 1983, and the chapter by H Böhme in this volume.

4 Prest A R, *How Much Subsidy*, Institute of Economic Affairs Research Monograph 32, London 1974.

5 Tzoannos J, *The Fiscal Regime for Shipowning Firms in the EEC*, Institute of Economic and Industrial Research, Athens 1980. See also Gardner B and Marlow P "An International Comparison of the Fiscal Treatment of Shipping", *Journal of Industrial Economics* Vol 31, 1983.

6 OECD *Maritime Transport 1984* , Paris 1985, Chapter 5.

7 Consultative Shipping Group. This consists of government representatives from Belgium, Denmark, Finland, France, Germany (FR), Italy, Japan, Netherlands, Norway, Portugal, Spain, Sweden, United Kingdom.

8 Lloyds List, *GCBS accuses the US of "blatant" protectionism* , 22 March 1984.

9 See Goss R O, *Studies in Maritime Economics* , Cambridge University Press, 1968, Chapter 2.

10 See *Lloyds List* , 22 February 1986.

11 E.g. the threat to ban Venezuelan vessels from US trades. (*Lloyds List 11 February 1985*), or the US moves that led the Philippines government to step down on its proposals for a bilateral cargo reservation policy (*Lloyds List 24 August 1984*)

12 Bredimas A E and Tzoannos J, In Search of a Common Shipping Policy for the EEC, *Journal of Common Market Studies*, Vol 20, 1981.

13 Ademuni-Odeke, The Brussels Package: The United Kingdom and the EEC. Respond to the Code, *Maritime Policy and Management*, Vol II, 1984.

14 The original draft Regulation proposed by the Commission is contained in COM(81) 423 of 13-10-81. OJ. No C282 of 5 November 1981.

15 *Council Report* MAR 37/14-9-84.

16 House of Lords Select Committee on the European Communities, 3rd Report, *Competition Policy: Shipping, HMSO 1983.*

17 See, Trade in the 1980s: Three Views. An Interview with Geza Feketekuty, Senior Assistant US Trade Representative, *Economic Impact*, Washington, D.C. US 1984.

18 Goss, R O. Social Costs, Transfer Payments and International Competition in Shipping. *Maritime Policy and Management*, Vol 12 1985.

19 See *Lloyds List*, 4 March 1986.

CHAPTER 4

SENSE AND SHIPPING POLICIES

Richard Goss

1 Introduction

The object of this chapter is to examine and discuss some ideas which have recently been put forward in the field of shipping policy by the General Council of British Shipping (GCBS, the employers' association) and by the European Economic Community. In large measure (though not entirely) the chapter follows and expands upon the memorandum of evidence put in autumn 1985 by my colleague Professor John King and myself to the Select Committee sitting under the chairmanship of the Rt Hon Viscount Rochdale to consider the EEC document *Towards a common transport policy: maritime transport* .[1]

The next section briefly reviews the well-known decline of the fleet of British owned and registered ships - by contrast with the growth of flags of convenience - and some studies of the subject. It concludes that they provide, in many ways, rather unsatisfactory explanations. Section 3 then examines the nature and relevance of the recent contentions of GCBS in the field of corporate taxation, finding that these tend to miss the main point but could nevertheless be much improved so as to be more acceptable to the Government. They could still provide much the same incentives to build ships - if these were thought to be desirable and effective; they would, however, be of less value to many of those who, ultimately, own much of British shipping.

Section 4 then summarises the argument of a recently published paper to the effect that British, and much other European, shipping is subject to other financial burdens, which appear as costs to the ship operators but which are transfer payments in economic terms. These, it is argued, are currently much more important, in sum and effect, than corporate taxes and represent one important reason for the simultaneous decline of British shipping and the rise of flags to which these imports do not apply.

Section 5 then examines an important argument missing from the above mentioned EEC document, namely the effects on such an unprotected sector as shipping of the levels of protection accorded to such other European activities as agriculture and manufacturing. The argument is essentially 'Euclidean' [2] rather than empirical in nature and anyone who finds it unconvincing (or inconvenient) is welcome to undertake the research necessary for its refutation. Should, however, anyone have the opposite reaction then the policy implications are significant, as is the range of options.

Section 6 of this chapter discusses the possibilities of dealing with flag discrimination in shipping along the lines of a maritime equivalent of the General Agreement on Tariffs and Trade (GATT). Section 7 draws the threads of the argument together, with some suggestions for the reconsideration of existing policies in Britain and in the EEC.

Overall, therefore, this chapter analyses various problems of shipping policy separately, since that is the only way in which they can be comprehended. The problems analysed are those that fall under the domain of public policy and are not strictly the business of the firms involved in the sector.

2 The decline of British shipping

In the year 1900 some 45% of the gross tonnage of world shipping over 100 tons was registered in Britain. By the middle of 1984 it was less than 4% and by deadweight tonnes it was not very different. True, the quality had risen very considerably between these years; and it is also true that the 4% is, on average, of a higher quality than most of the other 96% - but no one denies that there has been a substantial and disturbing decline.

At the same time there has been a considerable growth of the tonnages registered in other countries, notably the 'free registry' flags of Liberia and Panama, usually referred to as 'flags of convenience' or FOCs. Only a few of those ships are beneficially owned, operated or manned in those countries, though there are numbers of agencies and nominal companies which may have legal ownership and sometimes appear to manage them. All important decisions are, however, taken elsewhere.

As usual with examples of economic growth, there is supply and there is demand. The supply of these flags - and there are others - comes from the desire of their governments to earn revenue from the tonnage or registration dues that they levy. Such an activity is attractive to small countries and especially to those with little in the way of other exports. Despite a fairly high level of competition between them they earn a great deal from these dues; to describe these flags as 'tax-free' is thus to use a very restrictive definition of the term 'tax'; though of course it is true that the taxes paid are not related to the level of profits.

The demand comes from ship operators seeking a combination of low costs and freedom. The low costs stem largely from certain savings on the crew which are dealt with in section 4 below. The freedom, however, takes two forms. The first is a greater degree of flexibility in safety regulation. It is probably an over-simplification to treat this as if it were always a euphemism for lower safety standards, for many FOC owners have standards as high as those anywhere. It is true that the average loss records of FOC fleets are rather poor, but this is because they contain some quite different owners, whose idea of cutting costs is to cut safety and who, by one means or another, get away with it to a disturbing extent. This is not mainly an economic problem and I will not deal with it further. The freedom that is important in this context lies in the ability to adopt more flexible approaches to safety problems; producing, that is to say, solutions which are more cost-effective, even though they may not comply with the detailed rules of some national officials.

The second form of freedom is wider and is derived from the smallness of the FOC countries, both absolutely and in terms of international trade. For, suppose that they wished to impose taxes on the ships they have registered - or to do anything else that their owners dislike - then they have no means of enforcing it. Thus these FOC countries may have wars, revolutions and changes in shipping policies, but all may be avoided by re-registering the ships under some other flag, probably a competing FOC. Thus, FOC registration offers the double attractions of reduced costs with reduced uncertainty. By contrast, the registration of ships in Britain offers no such attractions.

The decline of British shipping has been studied in two well-known publications: the largely historical work by Sturmey [3] and a more descriptive and policy-oriented Report by an official committee [4]. Both examined the standard contentions of the spokesmen for the British shipping industry, namely that the decline of British shipping was caused by unfair competition from foreigners, mainly by flag discrimination but also by way of FOC fleets and subsidies. Both found that the quantitative effects of flag discrimination - the proportion of world seaborne trade affected by it - were modest, leaving plenty of other trade in which to engage. With the aid of detailed studies of British and other corporate taxation regimes (these are further discussed below in the next section) it concluded that British shipping ought not to suffer any disadvantages; and, when it came to the point, no one really pretended that subsidies could be a significant factor, since they were paid mainly to high-cost fleets such as those of the USA and then only so as to equalise their costs with ours. They also imposed various bureaucratic procedures and rigidity on the recipients. None of these factors seemed to have limited the growth of certain other fleets, for example that of Norway, and some were inconsistent with each other. For example, if flag discrimination limits British shipping, why does it not limit the growth of FOC fleets? Why are they not hit by subsidies? If flag discrimination hits British cargo liners, why not build other ship types?

This left Dr Sturmey, for one, concluding that the reasons for the decline of British shipping lay in its own backyard. It lay, he thought in such internal factors as the neglect of tankers when they were profitable; in persisting with steam propulsion when diesel was becoming better; in a general failure to build ships that were as fast and as big as their competitors; in the failure to use standard designs (he may have meant series production, a distinct concept which leads to much greater economies); in continuing to build in British yards when others were more efficient; and in continuing to blame others for its misfortunes. [5] This was doubtless exacerbated by a failure to make use of the highly favourable shipbuilding loans which were becoming available through export credit and similar schemes. Sometimes, too, there had been excessive nepotism.

A subsequent essay, vividly entitled 'Blood on the Thames' [6] records some of the vituperation Sturmey received. One writer told him that, although he had not read the book he felt that it ought never to have been written; many others joined in attacking him and on a variety of grounds. Yet the industry never published any well-considered response.

One can well understand the attacks, for each of his criticisms amounts to an accusation of bad judgement or management and is likely, therefore, to be resented by those accused. The absence of any serious response might be taken as confirming another of Dr Sturmey's comments, namely that the British shipping industry lacked any strategic machinery for collecting and making available information for the acceptance of its persistent complaints. In fact it persistently underestimated the quality of its audience, both in the public and among the policy-makers in the Ministry of Transport. (To this day, it is often embarrassing to have to explain some of their more naive statements to students.)

The Rochdale Report was couched in terms so understated that, misunderstanding, *The Times* said [7] that 'The report might almost have been written by the industry itself'; a sentiment which the industry itself would hardly have accepted. Nevertheless, though expressed quite differently [8] the Rochdale Report came to much the same conclusions as Dr Sturmey. In respect of flag discrimination, for example, (which the industry had represented as one of its most serious and intractable problems) [9] they concluded that:

> "only a small proportion of seaborne trade is affected by it at present ... (and) we do not consider that its total impact on British shipping has, so far, been particularly serious." [10]

They went on to say that it had serious effects on particular trades and was a danger for the future.

They also went on to make a large number of recommendations, most of which have been quietly watered down or allowed to gather dust; only one of these is of concern to the analysis here and this is covered in section 6 below. Nevertheless, though the Rochdale Report made few direct accusations of inefficient management, this - or some form of X-inefficiency - remains as the residual explanation. Naturally, it was correspondingly resented by the shipping industry.

Yet, it may be asked, is management so much more inefficient in British shipping than in other industries? Is the capacity for technical adaption less and, if so, why? Is the extent of nepotism greater? Are industrial relations worse?

These are all questions which are difficult to answer. It does not appear that there is any special reason to believe that management in British shipping is very different in quality from that elsewhere in the UK. Of course it varies much between one firm and another. On the contrary, the real problems lie in two other fields, which are dealt with in sections 4 and 5 below. First, however, it is necessary to discuss the contentions of the General Council of British Shipping to the effect that much harm is being caused by the changes introduced by the 1984 Finance Act.

3 The fiscal treatment of British shipping

In order to consider and compare the fiscal treatment of an economic activity in different countries and at different times it is necessary to take into account such factors as the rate of tax and any allowances which may be set against it.

Because these may well take effect at different times it is necessary to discount these to present values. Only then, and with appropriate assumptions on profitability and the rate of inflation can overall comparisons be made. [11] Though some of the formulations may appear complex the principles and applications are simple. For ease of presentation it is usual to express the present values per £1m of investment in ships.

Such techniques are consistent with standard investment appraisal methods and are, by now, sufficiently well-known for them to be employed within many British and foreign shipping companies and their advisers, as well as within the Government Departments concerned. They have also been employed by EEC, OECD and certain foreign governments, often using UWIST staff as consultants.

Both before and after the Budget of 1984 there were three fiscal effects applied to shipping companies liable for taxation in Britain: the corporation tax itself, levied at a specific percentage on the whole of the company's taxable profit; the tax allowances for depreciation and for the payment of interest. By using appropriate formulas, the present values of these may be calculated for any given rates of profit and inflation. The rate of profit is conventionally and conveniently expressed as an internal, or discounted cash flow, rate of return broadly equivalent to the 'yield to maturity' used when buying bonds. It is not equivalent to the accountancy concept of 'rate of return on capital employed', nor could it be, since that is bound to vary over the ship's life as depreciation accumulates and book values decline. A ship's life of 20 years is assumed and we shall use the government's test discount rate of 5% pa in real terms to produce our present values, pre-tax profit rates of 5%, 10%, 15% and 20%, all expressed in real terms, and inflation rates of zero and 5%. We shall also assume that the ship is financed on standard 'export credit' terms of 8% interest charged on the outstanding balance and with the loan repayable over 8 years. For our purposes, modest changes in such assumptions generally produce insignificant effects; they can, in any case, be explored by sensitivity analysis.

Before the 1984 Finance Act [12] the fiscal regime for ships owned by British-resident companies was fairly simple. There was a 'depreciation' allowance of 100% in the first year of a ship's operation, interest was deductible and the remainder was subject to corporation tax at 52%. The first year allowance could, of course, be carried forward by way of tax losses, but it was obviously more valuable to use it as soon as possible and a number of arrangements were made in order to achieve this. All of these take advantage of the fact that the tax is levied on the company and not on the ship - which, despite the existence of legal actions *in rem* - is not a legal person. They included company mergers, for example between property companies with good profits and small tax allowances and shipping companies in opposite circumstances; and leases by which the ownership of the vessel was taken by another company, for example a bank, which was able to take full and immediate advantage of the 100% allowance, the benefits of which were duly shared between the partners to the deal. Such arrangements were quite legal and highly respectable companies entered into them. They were sometimes termed 'tax-efficient', though the Treasury may have viewed them differently. So may the rest of us when we have to make up the lost revenue.

The 1984 Finance Act introduced a general series of reforms into corporate taxation, the general philosophy of which was that the Government no longer believed that favourable tax allowances stimulated overall investment; rather that, when combined with the tax-deductability of interest payments, they would distort the pattern of company investment, for example by inducing investments which produced less advantage to the economy as a whole than to the company making them. [13]

The changes involved the substitution of a 25% reducing balance depreciation allowance for the 100% first year allowance and reducing the rate of corporation tax (eventually) to 35%. The tax-deductability of interest was continued but its value was reduced by the fall in the rate of corporation tax. Obviously, this series of changes has the effect of making companies better off at the higher rate of profit and worse off at the lower ones.

This is demonstrated in table 4.1 on pages 82-83. Part A of this presents the results of the calculations specified above at zero inflation and part B with 5% inflation. The left-hand side of each part shows the pre-1984 regime and the right-hand side the current one. The three lines represent respectively the present values of the tax allowances for depreciation and interest, which do not vary with profitability, and the present value of the corporation tax, which does. The columns represent the four different levels of pre-tax profitability specified above. In the 'total' line plus signs represent net fiscal incentives to invest and minus signs net fiscal disincentives.

The results are summarised in table 4.2, which shows the differences between the old and new regimes for each of the 8 sets of circumstances considered in table 4.1. It will be seen that the change has made investments in ships less attractive at the two lower rates of profit but more attractive at the higher ones. This, of course, is consistent with the Chancellor's announced intentions but anyone in shipping may well retort that it is a long time since they saw rates of return like 15% or 20% *in real terms*. The present Chancellor, in turn, might respond that he is trying to induce high-profit investment and not the reverse. [14]

GCBS, perhaps wisely realising that the 100% first year allowances had gone forever, petitioned the Chancellor for a '50% ship allowance' to be additional to the 25% allowed by the reducing balance system in the first year, making 75% in the first year. Again, of course, it is the immediacy of the allowance which gives this its value. They claimed [15] that:

> "from the Government's point of view the cost of such allowances is limited to a deferment of tax receipts ... (and) without new investment there will be little tax revenue to collect."

These views seem to ignore the widespread existence of the 'tax-efficient' arrangement described above (which had numerous precedents under the previous regimes involving investment grants and investment allowances) all of which are well-known. They also seem to ignore the international ramifications of this industry, which permit tax allowances, like investment grants, to be exported to the permanent disadvantage of the country giving them. On a wider scale they seem to ignore the Chancellor's principal objectives.

But there is more to it than this. For a change in a fiscal regime applied to any particular economic activity or industry is likely to have both 'income' and 'substitution' effects, just as with the effects on consumers of a change in a price. Thus, those who were doing it already will be made better off by such measures as GCBS advocates, just as if they had acquired bigger incomes; because the activity has become more attractive it will also be done on a larger scale, both by those doing it already and by newcomers. The first may be termed the income effect and the second the substitution effect, again consistently with conventional consumer theory. Both these effects are commonly found in, for example, agricultural subsidy programmes and one common problem for governments seeking to expand some activity is how to concentrate effects on the substitution and to minimise the income effect. (Interested parties, however, often seek to do the opposite, since that is how they may benefit most from the least effort.)

Yet there is growing evidence in shipping that the effects of changes in fiscal regimes appear more in income than in the substitution effects. This evidence lies partly in the UWIST study cited above [16] and partly in a recent study by the British Maritime Charitable Foundation [17]. In the former, repeated attempts were made to correlate changes in the fiscal regime with those of the British fleet and in many ways. Many attempts were made because the statistics were not all that we would have wished: and no such relationship could be found except with investment grants when, it is generally agreed, many foreigners registered their ships through British-resident companies and, correspondingly, there was much undesirable exporting of our taxpayers' money. In the BMCF study, where former and actual shipowning companies were asked why their ships had gone overseas, the replies rarely indicated fiscal effects as being among the more important causes.

This may be thought surprising since British and foreign ships are generally excellent substitutes for one another, both in the short and in the long run and the mobility of at least some owners is well-known. But, certainly, there is no evidence for there being any high cross-elasticities of supply with respect to the fiscal treatment: yet it is, of course, implicit in all the GCBS arguments.

The claim that tax allowances represent no more than the postponement of tax could become valid if the Chancellor were persuaded to regard shipping as a special activity and to construct a 'ring-fence' around specialist shipping companies, as with the North Sea oil sector. No doubt some complications would ensue, for example in ensuring that transactions with other companies (possibly in the same ultimate ownership) were at arm's length; these difficulties are enhanced by the development of multi-modal transport systems; but it could certainly be explored and such a suggestion would greatly strengthen the case presented by GCBS. To the extent, of course, that some of the benefits would otherwise be immediate and permanent, such owners of British ships as banks would be worse off by such a scheme; and that may be why GCBS have not yet advocated it.

Finally, the main problem for British shipping is not that it has too much of its cash flow taken by the Government; nor, as table 4.1 reveals, that the fiscal incentive to invest in new ships is penal at the low real rates of return that the industry usually claims to achieve: it is simply that British shipping cannot make sufficient gross profits to survive. This is partly because of the present slump but it seems that a long-standing reason lies in some little-explored aspects of crew costs.

4 Non-wage Crew Costs in Shipping

As stated in the introductory section, this chapter is not concerned with such temporary events as the current shipping slump: but rather with what seem likely to be one of the permanent effects of it as far as crew costs are concerned. For, though it is generally true that shipping is a capital-intensive rather than a labour-intensive activity, in a time of slump crew costs must have a greater proportionate significance. Moreover, if many other ship operating costs are determined in international competition, why should this one be exempt?

There has, therefore, been much development in recent years of crewing agencies who specialise in the supply of cheap - they would say cost-efficient - crews, often from developing countries with large populations and modest accumulations of capital. It is not difficult to learn most of a sailor's skills, nor does it require much education or intelligence - as any sailor learns from his own seamanship instructors. Some education - literacy and numeracy - is needed for officers, but the technical training can easily be acquired, often in the schools of developed countries. None of this is remarkable and, with higher education, one can observe from his own students that there is nothing they cannot learn as well as British students.

Most of these crewing agencies - and many British ship operators - would add: "and they do it cheaper". Discussions of this subject often concentrate on the obvious subjects of wage rates (together with leave and other conditions of service) and productivity - for which manning scales are often used as a rather inferior proxy.[18] There is, however, a third component, consisting of what are sometimes termed non-wage costs and which is usually ignored or not separately identified. This is why it is seriously underestimated.

These non-wage costs are often considered to consist of the social security, pension scheme and other costs which are explicitly incurred by a British employer, whether in shipping or elsewhere. Certainly, these are both substantial and easily avoided by employing seamen from abroad: but my contention is that *the existence of international competition in shipping, and particularly that of the FOC fleets, effectively shifts on to the employer the burden of the deductions from the seamen's pay for similar purposes and for income tax.*[19]

Suppose a British sailor to have a nominal wage of £100; it is then demonstrable that he will cost his employer about £125 but will receive a take-home pay of some £77. An FOC employer, who may care little for any such schemes may then pay (say) £85, with the extra £8 going to compensate for the absence of a pension scheme. The British employer thus has wage costs almost 50% higher; moreover, since he will have to pay all this out in cash he will certainly regard it as part of his operating costs: yet, in economic terms, most of it represents transfer payments, ie payments not willingly made for specific goods or services received in return.

We have thus shown that, typically, a large and important crew cost advantage may go to the FOC operator. It is, therefore, hardly surprising that the ingenious British ship operator follows suit, often transferring the registry of his ships abroad or using an off-shore crewing agency which may re-employ the same British seamen at much lower cost, partly because they need not experience deductions for social security nor, if they time their own visits to Britain appropriately and have no home here, income tax either.[20] Naturally, in view of this, his wage rates are adjusted downwards and he is expected to make his own pension arrangements.[21] Numbers of brokers have sprung up to help him in this and their advertisements may be found in the appropriate journals.

A significant literature on non-wage labour costs has grown up in recent years.[22] Very little of it, as yet, refers to seafarers, but there is little doubt that the subject is of increasing significance; and especially in shipping because of FOC competition. Certainly, in discussing the overall level of crew costs, it needs to be considered separately from wages and productivity.

Of greater practical importance, however, is the question of what, if anything, might be done about it. As noted above, employers increasingly avoid these costs; it follows that *to relieve shipping of these burdens need not alter the public sector borrowing requirement.* Arrangements could thus be made for relief of social security and income tax for British-resident seafarers working for British shipping companies with only insignificant repercussions on the aggregate balances of central government revenue and spending. There might also be some economies in administration, which is currently quite remarkably complex and expensive to administer.

There is no real difficulty in making special arrangements to overcome such problems; many OECD member countries already do so in respect of personal income tax or social security deductions. On grounds quite different from those argued above and to minimal extents, the UK has done so for both of these, so precedents exist.[23]

5 "Free Trade" for shipping

The recent EEC approach is an extremely important statement setting out a coherent series of policies on a wide variety of maritime topics - though it quite ignores that dealt with in the previous section. It generally adopts a "free trade" position in respect of shipping, consistent with the theory of comparative advantage. Thus, given each nation's endowments, if we all specialise in those activities to which we are best suited, exchanging the resulting outputs of goods and services whenever it is advantageous to do so, then we will all be better off. There has been much discussion of this theory since it was first developed and there is wide agreement amongst economists that there are few exceptions to it. Nor, are they relevant to shipping.

It is argued here however, that these ideas are of general validity and that it is wrong to apply them to shipping alone.

Were this a world generally characterised by free trade then it would be easy to demonstrate that, the few exceptions apart, a deviation from free trade in respect of any given industry would tend to lessen economic welfare; and conversely that the correction of such an error would improve matters. This may not be true for everyone, since the gains and losses are likely to be unevenly distributed, but it can be shown that those who gain could compensate those who lost, while still remaining better off themselves. The fact is, however, that we do not live in such a world and the EEC is itself highly protectionist.

This is so in respect of most agricultural products, where the Common Agricultural Policy maintains supports for European farmers which reduce imports of such competing goods as sugar-cane, butter and lamb, thus raising internal prices, benefiting EEC producers and harming both its consumers and competing producers from outside. Such policies regularly (and predictably) lead to embarrassingly large surpluses which are difficult to disperse without undermining the original policy. In general, protection is also applied to manufactures, in aid of whose producers tariff walls surround the EEC. Should these be overcome on a large scale they are often supplemented by import limitation agreements having much the same effect. (Textiles and Japanese cars are examples). The exporters of such goods are often willing to agree to these schemes because their prices also rise and the increased profit per unit sold may well be sufficient to compensate them for the reduced volume.

If, however, we have an economy surrounded by a protectionist wall which is markedly uneven - in the sense that some industries are protected much more than others - then resources will be attracted into these industries which are protected and away from those which are not.[24] If, therefore, we are to have substantial levels of protection for many industries then it is desirable to consider any particular industry in the light of this and not in isolation. This is precisely what the EEC paper fails to do. To be specific, one of the many reasons why European shipping is contracting and suffering is that European farmers may expand and benefit.

Whatever we think of them, however, the Common Agricultural Policy and high levels of protection for manufacturing are fundamental in EEC practices. Nor is this the occasion to discuss them. For all practical purposes it is therefore necessary to take them as given and to consider appropriate policies for shipping in the light of this. This is what is done in the rest of this section; but it must be stressed that this argument is separate from and additional to that in the preceding section.

Shipping is an industry which, of necessity, operates internationally, selling its services in many countries and, correspondingly, buying such factors of production as insurance, fuel and the ships themselves on a world market. For most traditional flags, such as those of Britain and the other EEC members, trade unions, custom and common language needs, if not governments, have generally ensured that most of the crew usually have much the same nationality as the ship. Yet, in order to attract people to work in shipping for any length of time (i.e. in order to become trained and skilled) they must be provided with net advantages (including pay, leave, other fringe benefits and living conditions) broadly equivalent to those which they could obtain in the protected industries.

We thus have a position in which shipping employers will negotiate with seafarers' trade unions, claiming that low profits, essentially derived from low-wage competition from elsewhere, prevent them from increasing their own wage rates. The trade unions will probably respond by seeking increases in their members' pay etc. consistent with the rewards obtainable elsewhere in their homelands. If the trade union leaders are militant or if the employers enjoy fighting such people then the proceedings may become heated and to no good effect.

For, in such circumstances, both parties are right and the conflict between them, arising essentially from the inconsistency of government policies, can be removed in only two ways: by modifying those inconsistencies; or by a large-scale decline of the unprotected industry.

Part of the value of the EEC document lies in its full exposition of the nature and assumptions behind existing and proposed future shipping policies. Another part lies in the absence of any equivalent earlier documents from successive British governments. For there has never been a Green or a White Paper on UK shipping policies; the only official document is the Rochdale Report.

There has, indeed, been a general lack of public discussion on this subject - even though bodies like the Greenwich Forum and the British Maritime League have tried to put matters right. Because of this, and because governments neither understand nor, therefore, attempt to remove the inconsistencies, the resulting policies have contributed to just such a decline in British shipping.

Other countries, especially in the developing world, see the situation differently. They see their economic development, and especially their exports, limited by the protectionist policies of the developed world. They see that shipping is one of the relatively few industries to which entry is easy and in which protection currently extends to only modest proportions of output. Thus they sometimes decide to enter it; sometimes they bring their own forms of protection to it, seeing these as balancing the protectionism that we apply not to shipping but to agriculture and manufacturing. For developed countries then to lecture them on the evils of their ways, publicly or through diplomatic approaches - or to offer aid instead of trade - is worse than being patronising: it is analytically wrong; and from our point of view as well as theirs.

6 Action on flag discrimination

Whether flag discrimination has been an important reason for the decline of British shipping or not (a topic examined in section 2 above) and whatever the arguments about protectionism for other industries, there is little doubt that it causes various forms of inefficiency in shipping. It tends to substitute less efficient ships - those which need protection -

for those which do not; it tends to inhibit multilateral voyaging between groups of countries; and, because the viability of some shipping companies depends on it, their managers will devote more of their time to ensuring that they have good relations with the regulators and less to serving their customers. All this tends to raise the real cost of sea transport; and to a trading nation like Britain, this matters a great deal. (Much of this argument is complicated by the existence and practices of shipping conferences, but space does not permit a full consideration of that subject.)

Yet, even if flag discrimination be a form of sin, experience shows that it is fanciful to suppose that it can be eradicated. Can it, then, be limited?

The EEC Regulations refer, at many points, to the securing of non-discrimination or to the modification of discriminatory arrangements between EEC members, or other members of OECD, usually on a reciprocal basis. OECD is, of course, known as 'the rich man's club' though, because its membership reflects relative degrees of development when it was formed (does the UN really have no way of revising these matters?) it excludes several countries, for example Saudi Arabia, Kuwait and Singapore which have become quite rich, either absolutely or *per capita*. It also excludes all developing countries, whatever their policies on this and other matters and all communist countries.

Few EEC or OECD members practice flag discrimination on a significant scale: the main threat comes from developing countries with maritime ambitions, often using the UNCTAD Code of Conduct for Shipping Conferences as a justification - even though its provisions are purely permissive. The arrangements envisaged in the EEC Regulations are, therefore, largely aimed at the wrong target. (The main exception is the USA; but many years of experience suggests that the EEC Regulations are unlikely to work in that context, for the only thing likely to be effective is a change of heart by the USA.) By excluding the whole of what has been labelled 'the developing world' the EEC approach thus tends to perpetuate a division which is increasingly anachronistic as well as ineffective.

This is (most likely unintentionally) exacerbated by the tone in which the original Commission proposal was written.

The developing countries are treated as if they were homogeneous, which they are not and for a variety of reasons. By their omission from the essentially co-operative arrangements proposed for EEC and OECD members they are being treated as being essentially different, almost like potential enemies. Much of this, surely, must have spilt over from the controversial atmosphere at the meetings of the Shipping Committee of UNCTAD, which rapidly degenerate into negotiations between groups of countries organised according to whether they have been labelled 'developing', 'socialist' or 'developed market economy'. (Cross-membership is virtually forbidden, even when nations are the closest of friends on other matters: and some very curious bed-fellows appear in the same groups.)

If developed maritime countries wish to limit the spread of flag discrimination mainly from and by certain developing countries then surely it is necessary to treat all nations as if they were alike? Nor is it sufficient for reciprocal arrangements to be made under the third (or any other) Lomé Convention for that excludes much of Asia and all of Central and South America. The baby may not have been thrown out with the bathwater - since that implies action more vigorous than we are accustomed to seeing from Brussels: but it is an unnecessarily small and inadequate baby that remains.

It would, however, be possible for the EEC - or any member, such as Britain - to propose the creation of a new treaty or convention on flag discrimination. Essentially, it might have three rules:

1 No member would be allowed to exercise flag discrimination against any other member, though he would be free to do so against non-members;

2 Special arrangements would exist for members to discriminate collectively against non-members, or misbehaving members, whenever they agreed to do so;

3 Membership would be open to all nations.

This would allow any nation which wished to build up a modest shipping industry, for example for its perceived strategic or other reasons, be they good or bad, to do so; it merely refrains from joining the treaty and exercises some degree of flag discrimination. It can, however, do this only within the trades to and from its own shores: clearly it cannot do so in cross-trades.

But, given the practical complexity of modern shipping operations, cross-trade operations become a necessity as soon as quite a modest scale of operations is exceeded. Were such a nation to attempt to enter cross-trades then two effects are likely to become apparent. First, if its ships are genuinely less efficient, their unprofitability will be demonstrated. Secondly, no matter what agreements they may have made with their trading partners in their direct trades, they are likely to experience the full force of discrimination against them. If their ships are not less efficient then they will learn that they do not need the discrimination (except, perhaps, to force a recalcitrant conference to accept their line) and if their own discriminatory measures are modest then the reaction against them may be similarly muted. Nor, except in extreme circumstances, is the collective discrimination clause likely to be invoked; if only because non-maritime matters will usually be involved as well.

Thus, while any non-maritime nation could build up a fleet by means of flag discrimination, it could never become a major maritime power. It would be better if the EEC were to propose such essentially multilateral arrangements and upon bases of equality: not those which, when considered together with those discussed in the earlier sections of this chapter, resemble the rather tattered remnants of colonialism.

7 Conclusions

A number of threads have been followed in the preceding sections of this chapter and it is now time to pull together to see whether they are woven into any kind of logical pattern.

First, and most importantly, Sturmey was incorrect in reaching the conclusion that the decline of British shipping stemmed from internal rather than, as the industry would have us believe, from external causes. True, the external causes are not those publicised by GCBS but that is another problem.

In particular, this chapter identified a series of impositions upon British ship operators which have the effect of greatly increasing their crew costs but which are actually transfer payments. FOC competition is important and growing. In FOC fleets such burdens appear neither upon the employer nor upon the crew. Therefore the effective burden on the British employer includes both those payments for which he is explicitly responsible and also those made by deduction from the crews' wages. When quantified on this basis the burden is seen to be considerable and this probably acounts for the extent to which British ship operators are currently rearranging their affairs so as to dodge such payments. (The only mystery is that they did not start sooner.) Similar burdens exist in many other traditional maritime powers: though some of them modify these burdens substantially.

Secondly, it has been argued that the levels of protection accorded to other British and European industries are bound to (indeed, are intended to) make these more profitable than those sectors, like shipping, which remain unprotected. It should be noted that Sturmey himself argued along the same lines [25], saying (of the inter-war years):

> "In addition to reducing the volume of world sea-borne trade, tariffs had the effect of raising prices and wages in the protected countries making it more difficult for shipping enterprises in those countries to survive as they had to secure their resources in a protected market but to sell their services in an unprotected market."

He does not, however, carry this through to his discussions of post-war shipping, or to his conclusions; nor, writing in 1962, could he possibly have discussed the effects of the EEC as they are seen today. This, of course, is not to say that the internal factors which Sturmey brought out are unimportant: merely that there are important external factors as well.

GCBS, on the other hand, has concentrated exclusively on external factors *in shipping*, i.e. without making comparisons with the market situations of other industries. This attitude has led them to concentrate in recent years on corporate taxation, on which, as suggested above, their approach might benefit by some amendments. In this chapter it proved difficult to suggest any way in which their proposals might be reconciled with the Chancellor's basic strategy of reducing both tax-based incentives for investment and corporation tax so as to encourage investors to put their money where the pre-tax profits are greatest. Again, however, this is another problem.

The EEC approach thus seems to miss some points rather seriously, partly in failing to deal with non-wage costs and partly in failing to recognise the predictable effects of other EEC policies. After all, nations have limited resources; and if you deliberately expand some sectors then it is not much use complaining if others contract; and it is still less use blaming it on foreigners.

The EEC approach also seems, no doubt accidentally, to propose acting in an offensively patronising and discriminatory way towards those developing countries which practice, or are likely to practice, flag discrimination. Many of those most concerned will be automatically excluded from the arrangements proposed under the Lomé Convention. It may well be that this attitude has - accidentally - spilled over from the unfortunate controversies of the Shipping Committee of UNCTAD.

A new approach is therefore needed, so as to recognise the desires of nations to have some shipping of their own and to support it by limited degrees of flag discrimination, but to provide disincentives to their increasing it. This could be embodied in a treaty one of whose essential features would be that it would be open to all upon an equal basis.

Table 4.1: Present values of pre and post 1984 budget UK tax regimes, by item and profitability, per £1m invested in a ship, discounting at 5%

Pre 1984 100% 1st Year Allowance 52% Tax Rate				£1000s
	Rate of Return			
	5%	**10%**	**15%**	**20%**
Part A: Zero Inflation				
Item				
Depreciation	+483.3	483.3	483.3	483.3
Interest	+118.6	118.6	118.6	118.6
Tax	-483.0	-707.1	-962.4	-1237.1
Total	+118.9	-105.8	-360.5	-635.2
Part B: 5% Inflation				
Item				
Depreciation	+450.7	450.7	450.7	450.7
Interest	+96.1	96.1	96.1	96.1
Tax	-450.5	-660.9	-897.6	-1153.8
Total	+96.3	-113.2	-350.8	-607.0

Note: Signs follow cash flow conventions, + for incomings and - for outgoings.

(continued)

Table 4.1: (continued)

	Post 1984 25% Reducing Balance Allowance, 35% Tax Rate			£1000s
	Rate of Return			
	5%	**10%**	**15%**	**20%**
Part A: Zero Inflation				
Item				
Depreciation	+284.4	284.4	284.4	284.4
Interest	+79.8	79.8	79.8	79.8
Tax	-325.1	-476.3	-647.8	-832.7
Total	+39.1	-112.1	-283.6	-468.5
Part B: 5% Inflation				
Item				
Depreciation	+238.3	238.3	238.3	238.3
Interest	+64.7	64.7	64.7	64.7
Tax	-303.2	-444.2	-604.2	-776.6
Total	-0.2	-141.2	-302.2	-473.6

Note: Signs follow cash flow conventions, + for incomings and - for outgoings.

Table 4.2: Differences Between Present Values of Pre and Post 1984 UK Tax Regimes, Inflation Rate and Profitability

					£1000s
Rate of Inflation	**Tax System**		**Rate of Return**		
		5%	**10%**	**15%**	**20%**
0	Old	+118.9	-105.8	-360.5	-635.2
	New	+39.1	-112.1	-283.6	-468.5
	Difference	-79.9	-6.3	+77.3	+166.7
5%	Old	+96.3	-113.2	-350.8	-607.0
	New	-0.2	-141.2	-301.2	-473.6
	Difference	-96.5	-28.0	+49.6	+133.4

Source: Calculated from Table 4.1

NOTES AND REFERENCES

1 EEC: *Progress towards a common transport policy: maritime transport*, Brussels, 1985.

2 Imré Lakatos: *Philosophical Papers*, CUP, 1978, in Vol 2, Ch 2 argues that most economic argument is basically Euclidean in that it is derived from self-evident propositions with, admittedly, some empirical content. Generally, moreover, statistical and other evidence is insufficiently conclusive or anomalous for simple hypotheses ever to be falsified as Professor Sir Karl Popper would have us do. In real life, facts are frequently conflicting, statistics are at least occasionally dubious and hypotheses cannot be tested singly. This is not to say, however, that much work could not be done towards supporting the ideas suggested here and ensuring that they are, if not valid, at least plausible.

3 S G Sturmey: *British Shipping and World Competition*, Athlone Press, 1962.

4 *Report of the Committee of Inquiry into Shipping* (Rochdale Report), Cmnd 4337, HMSO 1970.

5 This passage paraphrases pp 400-1 of Sturmey, *op cit.*

6 The text, largely composed afterwards (sic), of a talk given to the Honourable Company of Master Mariners, 13 March 1963 and first published in S G Sturmey: *Shipping Economics: Collected Papers*, Macmillan 1975.

7 *The Times*, 7 May 1970, leader.

8 Lord Rochdale is one of Britain's leading Christian Scientists. This sect believe that, by affirming matter to be unreal and God or Mind to be infinite, the condition of the unreal material body may be brought into harmony with the real spiritual estate of man who, being the image of God, should manifest perfection. Thus, it is true knowledge alone that heals the sick and, for example, in its news coverage the *Christian Science Monitor* applies the theory that the mere discussion of evil invests it with seeming power: therefore one should concentrate on good. (See *Encyclopaedia Britannica* , Vol 4, p 562).

9 *Report*, para. 148.

10 *Ibid*, para. 161.

11 For an extended discussion of this subject in a maritime context, see B M Gardner, R O Goss and P B Marlow: 'Ship Finance and Fiscal Policy', a paper prepared for the Nautical Institute and published as a special issue of *Maritime Policy and Management*, Vol 11, No 3, July-September 1984.

12 For simplicity we omit such previous regimes as those involving investment grants, the transitional arrangements of the 1984 Budget, the complications of Advance Corporation Tax and some minor amendments to the post 1984 regime. We also assume that taxes are collected with an average lag of one and a half years. None of these affects the substance of the argument.

13 For a full expression of these ideas, see the Chancellor's Budget speech.

14 *British Shipping Review, 1985*, GCBS at pp 5-6 presents the case attractively.

15 *British Shipping Review, 1985* GCBS at pp 6.

16 Gardner, Goss and Marlow: *op. cit.*

17 R Woolrych (ed): *Why the Ships Went* , BMCF 1986.

18 The structure of the crew, e.g. the proportions of different grades, can be as important as the total number.

19 The subject, its causes and implications, is covered at greater length in a recent paper: "Social costs, transfer payments and international competition in shipping" in *Maritime Policy and Management*, Vol 12 No 2, pp 135-43.

20 If a rating does not wish to do so he no longer has to belong to a trade union, so he can save the subscription too. Many FOC operators choose not to recognise trade unions.

21 With commendable frankness much of this is described in *British Shipping Review 1985*, pp 7 - 8. For a specific example (BP Shipping) see Lloyds List 10 Jan 1986 *et seq*.

22 E.g. R A Hart: *The economics of non-wage labour costs*, Allen & Unwin, 1984 provides a thorough general introduction and much useful theory. The EEC publication *Non-wage labour costs, their importcnce and effects* is also informative.

23 *Maritime Transport, 1984*, OECD, Paris, gives a curiously bureaucratic view of the taxation arrangements for shipping in most OECD member countries - all description and no analysis. However, right at the end of the relevant chapter (pp 143-4) there is a brief section on the differences between the tax and social security treatment of seamen and other working citizens. The contrast between this and the forty three pages devoted to corporate taxation provides a fine example of the bias which often appears in discussions of international competition in shipping.

24 Of course, if the protected industry were declining anyway, then its decline would probably be slowed, stopped or reversed. Indeed, protection is often applied to declining industries. The point is that it would attract or retain more resources than it would have done without the protection. Correspondingly, if the unprotected industry were expanding anyway, then the protection of the other would tend to reduce its rate of growth rather than to induce any absolute decline.

25 Sturmey, *op.cit.* p 72

CHAPTER 5

THE QUALITY OF SHIPPING SERVICES: AN ECONOMIC APPRAISAL OF STRATEGIES FOR THE EUROPEAN LINER SHIPPING INDUSTRY

Mark Casson

1 Introduction

The European shipping industry has traditionally supplied international liner shipping services to a large part of the free world. It is not surprising therefore that a key element in the European Community's maritime policy has been to promote the liberalisation of trade in shipping services and to discourage governments from implementing restrictive policies such as cargo reservation. [1] Liberalisation is a concept that is construed very broadly in the context of the shipping industry, for it includes the freedom of shipping lines to control services on any given route through a conference system - a point that is developed further below.

The EEC's preference for liberalisation seems to reflect not merely a desire to defend the freedom of its member countries' lines to operate on existing routes, but also an implicit belief in the long run competitiveness of the European liner shipping industry. This accords with the view that, in a typical industry it is the nations which believe their firms to be competitively advantaged that tend to favour liberalisation, and the ones which believe their firms to be disadvantaged that tend to favour protectionism.

In defence of the EEC's attitude, it may be noted that during the last twenty years the entry of the Soviet and East European fleets, and of the state-owned fleets of some developing countries, has not fatally damaged European international market share. These new fleets are often badly managed, and use fairly old and small vessels (or new vessels of rather dated design). They do not seriously compete for the most valuable cargoes on the routes that carry the heaviest traffic (on this, see also Chapter 7).

This is not true of all new entrants to the liner industry, however. Fleets based in South East Asia - Hong Kong, Taiwan, Japan and Republic of Korea, for example - now pose a serious threat to the Europeans. In the past, the conference system may have helped to buffer established lines from the full force of new competition by providing the members with collective strength to fight outsiders, and with monopoly profits which could be tapped to 'buy off' successful entrants. The power of the conference system has diminished recently, however, particularly on routes dominated by container traffic. Some of the new entrants - such as the Evergreen round-the-world service - do not seem particularly keen to join conferences on a permanent basis on all the routes on which they operate. It may be that some new entrants now believe that they are sufficiently competitive to eliminate conference members *en bloc* through efficient, low-cost operations.

During the past decade European manufacturing industry has suffered seriously from South East Asian competition. Industries such as motor vehicles, consumer electrical goods and ball bearings, which in 1970 were major net exporters, were by 1980 protected against foreign competition by 'voluntary export restraints' and EEC anti-dumping rules. In some cases, a European industry has nearly been eliminated as an effective competitor in non-European markets. In other cases, European firms have fought back through restructuring and new investment. Restructuring, however, has meant not merely cost reductions, but also major improvements in product quality. It has been recognised that European manufacturers cannot, in the long run, compete on price alone with manufacturers based in low-wage newly industrialising countries. To avoid low-wage competition, successful European manufacturers have typically moved 'up market', producing high-value items which command a quality premium.

The threat from new sources of foreign competition in the shipping industry may well force European shipping firms to consider a similar strategy. So far, however,

European firms seem to have perceived the situation rather differently. They have typically alleged that foreign operators achieve low costs mainly through state subsidisation and/or the use of flags of convenience, and have used this argument to demand matching subsidies for themselves. [2]

Subsidisation and flags of convenience are not a complete explanation of the growth of foreign competition, however. It seems likely that on certain routes foreign operators also achieve low costs through effective management. Dismissing foreign competition as a product of subsidisation may be politically convenient, but it is also a recipe for complacency. The complacency may have been encouraged, in the past, by the security that the conference system afforded its members. Unless European lines enhance their competitiveness, however, they may in the future begin to find themselves competitively disadvantaged by efficient large scale entrants to their most valuable markets. A determined effort to enhance quality of service is one way to stay ahead of the competition. This strategy would also benefit the European economy as a whole. Overall economic efficiency is more likely to be promoted when managerial effort is channelled into improvements in efficiency than into lobbying for subsidies.

If this point is conceded then it is necessary to enquire what steps can be taken to enhance quality. But there is the problem that higher quality is typically associated with higher cost. How far should the pursuit of quality be carried? Section 2 discusses the valuation of quality in the market for shipping services. It analyses the way that demand factors - namely customer preferences - and supply factors - namely the technical options available to firms - interact to determine the equilibrium level of quality.

Section 3 examines in more detail the 'quality revolution' that is underway in many industries, and which is bound to impact, both directly and indirectly, on the shipping industry. Section 4 distinguishes two aspects of quality: production quality and transaction quality. Production quality is concerned with the logistical aspects of transporting cargoes, while transactions quality focuses on contractual relations between the customer (the shipper) and the supplier (the shipping firm).

The efficient supply of production quality is complicated by 'system effects', which are examined in detail in Section 5. Transaction qualities, by contrast, are readily controlled by individual firms. They have implications, though, for vertical integration and diversification in the firm's operations, which are examined in Section 6.

Section 7 considers alternative methods of dealing with system effects in the supply of production quality. It is argued that international regulation, effected through the licensing of schedules services, has much to recommend it as a substitute for the traditional conference system. The conclusions are summarised in Section 8. It is suggested that European lines should move 'up market' in respect of both production and transaction qualities, and that to enhance the production quality afforded by the European industry as a whole, a European-wide alternative to the conference system should be devised.

2 Trading off cost and quality

In most industries - including shipping - there is a trade-off between cost and quality. Under what circumstances then, should improvements be focused on quality enhancement rather than simply on cost reduction?

Economic theory suggests that there are two circumstances under which competitive strategy should be biased in this way. The first is when new more efficient techniques afford greater costs savings on high quality services than on low quality services. The second is when there is a shift in customer preferences towards a high quality service. These two cases are examined in turn below.

Suppose that quality can be measured by a bundle of characteristics such that it is meaningful to talk about there being 'more' or 'less' quality in one situation than in another. The measure of quality defined in this way is plotted along the horizontal axis in Figure 1 (p 124). Cost is measured on the vertical axis. Suppose that there is a fixed amount of traffic to be carried over a given route. If the combination of freight charges and quality of service is unacceptable to customers, the extra traffic will be switched to another route. Initially the route is monopolised by a conference dominated by European lines. The efficiency of the conference is indicated by the cost-quality trade-off $C_O C_O'$. The curve is drawn convex to the horizontal axis, indicating that as quality increased, additional quality is achieved at progressively higher cost: in other words, the marginal cost of quality increases with respect to the level of quality already attained.

Conference policy determines the quality of service and hence the total costs of the constituent lines. It is assumed that the conference maximises the combined profits that its members derive from the route. This means that the conference maximises the excess of total revenue over total cost.

It is assumed that all shippers have identical preferences with respect to quality, and that these are reflected in the way that the total revenue earned by the conference is governed by the quality of service it provides. This relation is indicated by the curve $R_O R_O'$. The slope of the curve at any point measures the marginal quality premium - the additional payment that shippers are willing to make for a marginal unit improvement in service. The marginal quality premium is assumed to be positive but to diminish continuously as the level of quality already achieved increases. This is reflected in the fact that $R_O R_O'$ is upward-sloping and concave to the horizontal axis.

The curve $R_O R_O'$ is only one of an entire family of such curves. The curve $R_1 R_1'$ affords a higher level of satisfaction than $R_O R_O'$ because it provides a given level of quality at a lower charge. It is assumed that the marginal quality premium is independent of the level of charges set by the conference: this means that each revenue curve corresponds to a vertical displacement of any one of the others. Thus the slope of $R_1 R_1'$ at any point is equal to the slope of $R_O R_O'$ at any point directly above it.

The profit-maximising strategy is to set quality of service at the point where the marginal quality premium, as measured by the slope of $R_O R_O'$, is equal to the marginal cost of quality, as measured by the slope of $C_O C_O'$. This condition determines the profit-maximising quality OQ_O, and generates profit equal to AoBo. The conference revenue is $Op_o{}^m$ and the cost of Oc_O.

Suppose now that the conference is challenged by an outsider of similar efficiency. Both suppliers now share the cost-quality trade-off C_0C_0'. Each supplier bids for the whole of the traffic, forcing down revenue to a point where they both break even. The actual outcome is the break even point on C_0C_0' which affords shippers maximum satisfaction, namely the point of tangency B_0 between C_0C_0' and R_0R_0'. Price is reduced to $OP_0{}^c = Oc_0$, but quality remains the same as before.

In this particular case there is no point in either of the parties attempting to meet the competition by increasing quality rather than reducing charges, because both the marginal cost of quality and the marginal quality premium are independent of the level of charges. Thus if the conference attempted to meet outsider competition by operating at D, say, which gives high quality for the same charges as before, its rival could win away the customers by choosing a break even point on the segment B_0D. Any such move, which involves cutting price at the expense of quality, would make the customers more satisfied.

Notice that, unlike the case where **quantity** is concerned, the change in market structure does not alter the amount of **quality** supplied.[3]

Suppose now that in an effort to fight foreign competition, members of the conference improve efficiency and thereby shift down the cost-quality trade-off to C_1C_1'. The figure illustrates the case where the greatest cost savings are achieved with the highest quality of service. The pricing of services by the conference must allow customers to achieve at least as much satisfaction from the new service as they could obtain from any service operated by the rival on a break-even basis. This restricts the conference to the revenue curve R_1R_1'.

Equating marginal quality premium to the new marginal cost of quality determines the optimal OQ_1, generating a profit for conference members of A_1B_1. The charge for the new conference service is Op_1 and its cost of supply is Oc_1. Notice that the charge Op_1 is actually higher than the charge set by the rival, $Op_c{}^0$. This reflects the fact that the conference is providing an enhanced quality of service for which the customers are willing to pay. The fact that the optimal quality is higher than before reflects the fact that the efficiency improvements have reduced the marginal cost of quality at each level of quality. Had the improvements *increased* the marginal cost of quality (whilst reducing the average cost of quality through a lump sum reduction of cost independent of quality) then the optimal level of quality would have *fallen* instead.

It must be emphasised that even in the case where the marginal cost of quality falls, the unrestricted pursuit of quality is not an effective strategic weapon. If, for example, the conference used its new methods to offer a level of quality in excess of OQ_1', the charges it would have to make in order to break even would be so high the customers would prefer to use the lower quality service of its rival instead. Efficiency leads to higher quality, therefore, because the improvements are biased towards higher quality and not because high quality is desirable whatever the cost.

Suppose now that there is a change in customer preferences, which raises the marginal quality premium at each level of quality. The situation is illustrated in Figure 2. The level of customer satisfaction associated with the curve R_1R_1' is now associated with the more steeply sloped curve R_1S. Consider first the implications for the rival firm that has experienced no improvement in efficiency. This firm will now afford a higher level of satisfaction to customers through break-even operation if it produces at G rather than B_O. At G the revenue curve R_2R_2', which represents a parallel displacement of R_1S, is tangent to the trade-off curve C_OC_O'.

The conference must offer customers at least as good a deal as they obtain at G. This means that R_2R_2' is the revenue curve facing the conference. The conference maximises profit when the slope of R_2R_2' is equal to the slope of the trade off C_1C_1' directly below it. This determines the optimal quality OQ_2, which exceeds the previous level of quality OQ_1. Notice that the conference substitutes in favour of high quality and higher cost just as its rival does. Because the conference's methods allow greater cost savings at higher quality, the substitution in favour of high quality raises conference profits from A_1B_1 to A_2B_2.

The moral of the exercise is fairly clear. Quality enhancement is an effective strategic weapon only under certain conditions. It can worsen the competitive position of a liner conference as well as improve it. The case for quality enhancement rests entirely upon the presence of either demand side effects or supply side effects which are biased towards greater quality.

3 The increasing demand for quality

It is the main thesis of this chapter that demand side effects are currently encouraging the pursuit of high quality in all industries, including shipping. The role of the supply side effects is rather ambivalent, at least so far as the shipping industry is concerned. On the one hand, it can be argued that containerisation affords the greatest costs savings in the provision of higher quality services. On the other hand, the improved technical ability to exploit economies of scale in vessel construction and operation has allowed significant economies which are independent of quality, and may have discouraged the pursuit of quality by encouraging operators to reduce the frequency of sailings in order to utilise larger vessels.

This section concentrates on examining the nature of the demand side effects. It is well known that the demand for the shipping of freight is a derived demand: this applies not only to the demand for the **quantity** of services but also to the demand for **quality.** The shipper's demand for quality stems from two considerations. First, high quality shipping enables him to provide a better overall quality of service to his own customers. Thus a manufacturer whose overseas customers demand prompt delivery will demand equally prompt service from the shipping line that handles his export trade. Secondly, high quality shipping reduces the shippers' own costs: for example, a high frequency of regular high-speed sailings reduces the shipper's inventory costs.

Consider first the attitudes of the shipper's customers. There is considerable evidence that individual consumers are becoming increasingly quality conscious. [4] This new awareness of quality has been a major source of problems and opportunities in manufacturing business during the last twenty-five years. Economic historians have often noted that rising real incomes cause consumers to become satiated in terms of quantities of familiar products, and to look instead for higher quality, and for more novel products. This seems to be particularly true of upwardly-mobile social groups, who regard their ability to recognise and appreciate quality as a hall-mark of their social achievement; quality snobbery is a major element in the demand for quality.

The related phenomenon of the demand for novelty is the basis for whole industries which are geared to the production of quickly obsolescent fashion goods. One aspect of the most recent generation of upwardly mobile groups, however, is an apparent disdain for certain kinds of mass-produced fashion goods. In some mature Western economies it has now become fashionable to regard novelty as being essentially superficial. There is also the fear that when products are genuinely novel they are unreliable, and therefore inconvenient and time-wasting for their owners. This more practical concern may reflect the increasing labour force participation by married women, which has led to greater emphasis on convenience and time-saving in household consumption activities. Consumers want products to be immediately available from stock, and to perform with total reliability once they have been purchased. The 'quality revolution' also has a cultural dimension. Intolerance of bad services seems to be becoming more acceptable socially, and customers who were once afraid to complain are becoming more vocal as a result. This may affect the shipping industry not only because it affects the manufacturers whose cargoes are carried by the shipping lines, but because the manufacturers themselves may be influenced by this more aggressive attitude in their own dealings with the shipping lines.

One of the ways in which progressive firms in mass production industries have responded to the 'quality revolution' is to redesign their products in multi-component form. [5] This leads to a simpler and more functional appearance of the product, which can be aesthetically quite appealing. It allows more versatility to be built into the product, and makes differentiation of models through the substitution of components fairly easy. It reduces servicing costs, and prolongs product life, by allowing repairs to be effected using interchangeable parts.

Multi-component design also reduces production costs by allowing the production of different components to be concentrated on different locations. Precision components, for example, can be produced in large quantities on automated production lines, whilst components where precision is less critical can be produced using more traditional labour-intensive methods. Production is organised along global lines, with different components being produced in different countries and shipped in bulk to assembly lines close to the major centres of final demand. This has led to a considerable increase in intermediate product trade. The prominent role taken by Japanese and US firms has meant that the growth of such trade has been concentrated in South East Asia and the Pacific, leading to a rapid increase in shipping requirements in this area.

Global rationalisation has meant that manufacturers now find the quality of shipping services to be a crucial factor in minimising their overall costs. Because of the large volume of components in transit between factory and assembly line, successful global rationalisation requires careful inventory management. Rationalised production is particularly vulnerable to disruptions to transportation since both component production and mass assembly afford considerable economies to continuity of flow. On both these counts, the reliability of shipping schedules assumes crucial importance to the rationalised producer.

In rationalised production systems, the quality of individual components is a crucial constraint on the overall quality of the finished product. Damage to components in transit to the assembly line must be avoided at all costs, lest the manufacturer's reputation be tarnished by undetected defects incorporated into the final product. In this respect, containerisation has made a substantial contribution to the quality of shipping services, since one of its main advantages is that it reduces the incidence of damage during transhipment.

If these related trends towards greater quality awareness amongst consumers, and global rationalisation amongst producers, continue, then it seems inevitable that the shipping industry will be caught up fully in the 'quality revolution'. The shipping industry is, however, an extremely complex industry, and it may therefore encounter special difficulties in responding to the situation. The next section of this chapter examines the nature of these problems, while subsequent sections discuss possible responses to them.

4 The dimensions of quality

When discussing quality of service it is important to distinguish between 'production qualities' and 'transactions qualities'. The production qualities of a service are those that reduce the customer's production costs, i.e. the cost associated with the physical transformation and movement of goods. Transactions qualities reduce the customer's transaction costs, i.e. his costs of setting up trade, negotiating, and enforcing contracts. The application of these concepts to the shipping industry is illustrated in the tables (pp 127-8). Table 5.1 lists the five production qualities that would seem to be most important where the shipping industry is concerned, whilst the five most important transactions qualities are listed in Table 5.2.

It is useful to begin by analysing production qualities in isolation from transaction qualities, and then to introduce transaction qualities once the significance of production qualities is fully understood. This is particularly important in the case of the shipping industry because of the complexity of the factors affecting production quality. It is not generally the case that a single type of quality-oriented investment will affect just a single aspect of quality of service in the industry. It might be suggested, for example, that the incidence of damage would be determined principally by the quality of stowage, and in particular requirements of the cargo. In this case, investment in vessel equipment would be linked in one-to-one correspondence with the incidence of damage. But, as already noted, the incidence of damage also depends crucially on the amount of transhipment involved, and this depends in turn upon the routing of vessels over the network of shipping lanes as a whole.

It could also be argued that the amount of space capacity available for last-minute consignments is in one-to-one correspondence with a particular quality-oriented investment. The amount of spare capacity provided on a voyage, it could be suggested, is determined principally by the size of vessel, and the principal effect of building a larger vessel is to increase the margin of spare capacity. For reasons that are well known, however, this is far too simplistic a view. The spare capacity available is determined by the size of the vessel *relative to the amount of traffic* , and the amount of traffic may vary over different legs of the voyage. Once the itinerary has been fixed, the supplies of capacity over different legs of the voyage are joint products determined by the size of the vessel, and they cannot be adjusted for variations in the level of traffic over different legs. These variations are governed by the pattern of consignments between different ports and the availability of other routes linking various pairs of ports along the route - including both direct and indirect routes between them. Thus the pattern of trade and the structure of inter-route competition affects capacity utilisation just as much as the size of the vessel.

The overall level of capacity utilisation on a voyage will depend upon the frequency of sailings on the route. If vessels of a given size depart every week then capacity utilisation will be much lower than if the service is fortnightly. This shows that inter-temporal competition on the same route also affects capacity utilisation.

Even if it were possible to control for factors influencing the demand for capacity, it would still be wrong to analyse investment in capacity in isolation, because size of vessel affects other aspects of quality of service besides the ability to accept last-minute consignments. When a vessel is following an itinerary that involves several intermediate ports of call, operating with a margin of spare capacity reduces the problems of stowing cargoes in the right order, thereby cutting stowage times and providing a quicker overall service to the customer.

These elements of inter-temporal and inter-route competition that affect capacity utilisation are a specific manifestation of the more general phenomenon of 'system effects' within the shipping industry. [6] These system effects become even more pronounced in respect of the third production quality, namely duration of transit. The total time that a consignment is in the hands of a shipping firm is the sum of five components:

(i) the time spent travelling at normal speed at sea;

(ii) the time used up by delays to the vessel in entering and leaving port;

(iii) the time spent at intermediate ports of call on the voyage (where more than one leg of the voyage is involved);

(iv) the time spent being loaded onto the vessel, and unloaded from it; and

(v) the time spent on the quayside awaiting transhipment from one vessel to another (where more than one voyage is involved).

Travelling time at sea is reduced if the vessel is speeded up, but beyond a certain point this leads to increased fuel costs. The advantage to the customer of saving time at sea will normally depend upon what proportion of total transit time is spent at sea, and this in turn depends upon the other four components of transit time.

Delays in entering and leaving port represent an important system effect. Assuming that the port has sufficient capacity to cope with the loading and unloading of consignments, congestion will be determined by the number of conflicting traffic movements in the vicinity of the port, which is related, first and foremost, to the number of vessels using the port in any given period. This in turn depends upon the number of routes originating at, passing through, or terminating at the port, and the frequency of the services on each route. For this reason, the choice of frequency on one route has spill-over effects on the quality of service on all other routes using the same port.

The time spent at intermediate ports of call depends upon whether the shipping system is based upon non-stop voyages supplied by feeder routes or on long voyages without feeders making many stops. The first approach minimises the time spent by consignments on-ship, but increases the use of transhipment, and hence raises the time spent on the quayside awaiting connections. Long voyages with many stops also create difficulties with stowing cargoes in the right order, so that as the number of calls is increased, the time allowed in each port has to be increased as well.

It is worth noting that where the duration of the voyage is concerned, there is a potential harmony of interest between the shipping firm and its customers, because a reduction in voyage times not only contributes to reducing overall transit times but also permits more effective utilisation of vessels. For a service of given frequency, the faster is the voyage, the smaller is the number of vessels required to maintain the service. The coincidence of interests is by no means complete, however. In the absence of quality considerations, the shipping firm would operate vessels more slowly than otherwise. It would also place greater reliance on non-stop voyages with feeder routes, since this would allow better utilisation of vessels. Customers would be obliged to accept non-stop voyages at speeds slower than they desire, and delays caused by transhipment that they would rather avoid.

In the absence of quality considerations, shipping firms would provide a minimal frequency of sailings on each route, using the largest vessels that can be accommodated in the ports. This is because of economies of scale to vessel size, which mean that in an extreme case, the entire system might be operated by a single vessel, taking on different routes in turn. In contrast to the previous case, there is a striking conflict here between the interests of the shipping firm, and the interests of the customer. The customer prefers a very high frequency of service on each route, so that there is always a service available at a time convenient for his own production operations.

5 Production quality on a network

The preceding rather general observations may be clarified by a simple example. Figure 3 illustrates an elementary triangular network of trade. The network is defined in terms of 'nodes', 'quays' and 'terminals'. A node is any point where two or more shipping lanes meet. [7] It is a potential point of congestion caused by conflicting traffic movements. It is assumed that all delays due to congestion occur at nodes rather than along the lanes - this is a reasonable assumption since the amount of traffic at a node will normally exceed the amount along any of the lanes leading to it.

Some nodes may exist where shipping lanes intersect out at sea, but in this simple example all the nodes lie in the proximity of quays where ships call for loading and unloading. A further simplifying assumption is that all ships passing through a node actually call at the quay - in a more general case it is quite possible that an express coastal voyage, say, might pass through a node near to a quay without stopping at the quay itself.

Not all ports provide terminal facilities such as routine inspection and maintenance. In this example, ports 1 and 3 provide terminal facilities but port 2 does not. The location of terminal facilities takes on particular importance when it is assumed that all vessels must eventually return to the terminal from which they set out. This assumption is rather strict, but may be justified on the grounds that maintenance is best effected by staff who are permanently located at one particular port and who have already become familiar with the vessel by doing previous maintenance work upon it.

Suppose, for example, that each vessel must return home after no more than four legs of a voyage. A restriction of this kind would arise if, say, there were equal distances between each pair of ports and each vessel required maintenance after a mileage (or time) interval equivalent to four trips between them. The restriction of the voyage to only four legs is unduly strict, of course, but in the context of an elementary triangular network the consequences of the assumption are not particularly serious. The main effect of the restriction is to limit the number of possible voyages. Allowing up to four legs to a voyage on a triangular network determines six permissible voyages from terminal 1 and six from terminal 3. The itineraries of these voyages, and the lanes that they use, are shown in Table 5.3. In each case, the first two of the voyages are 'out and back' shuttles to one other port. It should be noted in passing that two such voyages can be effected within the maintenance interval, and that by following a shuttle to one port with a shuttle to the other port it is possible, in principle, to provide through carriage of cargo between the two ports *via* the terminal at which the vessel is based; however, it is assumed for the sake of simplicity that no through working of this kind is in fact provided.

Some of the voyages pose more complex stowage problems than others. These stowage problems arise when the cargo is heterogeneous so that specific items - and not just a certain proportion of the cargo as a whole - have to be unloaded at each port. The shuttles have the simplest stowage properties, with the return loop voyages next in order, and round trips posing the greatest stowage difficulties. The stowage properties of the voyages starting from terminal 1 are examined in detail in Table 5.4, whilst the properties of voyages starting from terminal 3 are summarised in Table 5.5.

The triangular configuration of the network is much more general than may at first appear. This is because - in spite of the remarks made earlier the actual distances do not have to be equal, and the surface on which the nodes are placed does not have to be flat. Figure 4 illustrates various applications of the triangular network to shipping across the Atlantic, with the final application showing how Atlantic shipping may be viewed as part of a global network.

In deriving the list of voyages shown in Table 5.3, it was assumed that all ways of linking together the successive legs of a voyage were permissible. The application of the network can be extended further by relaxing this assumption. By imposing restrictions on the way that different lanes can be used as part of the same voyage, it is possible to model a multi-modal network, in which some of the lanes, say, are used by rail traffic, some by road traffic, and only a few by shipping. The same principle of restricting the sets of lanes from which a voyage can be constituted makes it possible to model situations where vessels used on one lane (say along a river estuary) cannot be used on another lane (say across the ocean) and *vice versa*. Finally, by eliminating the possibility of using certain lanes altogether, it is possible to create a linear network with or without an intermediate node as a special case of the triangular configuration. Figure 5a illustrates a multi-modal triangular network whilst Figure 5b illustrates a trunk route with feeder in which the network is linear and there are no through voyages over the two segments of the network.

The set of possible voyages, once determined, governs the number of different ways in which consignments can be routed. Each consignment can be routed either directly or indirectly. It is one of the characteristics of an unrestricted triangular network that the direct route between any two quays involves no intermediate nodes and that there is a unique indirect route which involves just one node. The relation between direct routes and indirect routes for an unrestricted triangular network is exhibited in Table 5.6, and the correspondence between the lane used by the direct route and the lanes used by the corresponding indirect route is shown in Table 5.7. The voyages that are available for indirect routing without transhipment are shown in Table 5.8. Using the information contained in Tables 5.6 - 5.8 it is possible to relate the size of vessel and frequency of sailing on each type of voyage to the amount of traffic that can be routed directly between any pair of ports, the amount that can be routed indirectly without transhipment, and the amount that must be transhipped at an intermediate port. The distribution of traffic between these three categories is a major influence on several aspects of quality of service - in particular the incidence of damage due to transhipment and the overall transit time.

The pressures that the different voyages exert on the various nodes is indicated in Table 5.9, together with the amount of storage time associated with each voyage. These factors govern the way that the frequencies of the different types of voyage impact upon both congestion at particular nodes, and the amount of time that the fleet as a whole is tied up at the quayside by stowage operations.

A formal model incorporating these effects is presented in the Appendix (pp 114-23). Using a model of this kind, the nature of the system effects described in section 3 can be made precise, and the trade-off between cost and quality of service can be derived. It can be shown that the trade-off between cost and quality of service at the level of the individual consignment is dependent upon the way these factors are traded off for other consignments which use other parts of the network. Thus the true trade-off between cost and quality occurs at the level of the system as a whole. Any attempt to trade off cost and quality for individual consignments without considering the system effects is almost certain to result in sub-optimisation.

6 Transaction qualities

Many of the transaction costs borne by the user of shipping services are incurred in finding a shipping firm that provides a suitable service, and in putting things right when the service occasionally goes wrong. [8]

An important feature of transaction costs is that they contain fixed components. In particular they contain a component that is independent of the number of different customers using a given service. In searching out a suitable shipping service, the customer is liable to replicate a great deal of detective work that has been done by other customers seeking a similar service. Since it is the nature of the liner trade that a relatively small number of suppliers service a large number of different customers, economic efficiency calls for the suppliers to take the initiative in reducing the customers' search costs by providing marketing services which all prospective customers can readily avail themselves of. The marketing services help the customer to make contact with the supplier, assimilate accurately the nature of the services supplied, assess what particular 'mix' of offerings he is interested in, and quickly conclude a deal. These points (and some others) are included under the headings A-C in Table 5.2. Because the saving to customers afforded by these services exceeds the cost to the suppliers in providing them, these marketing services reduce overall transaction costs in the provision of shipping services.

Another feature of transaction costs is that they contain a fixed element which is independent of the actual amount of traffic that is involved. Once a suitable shipping firm has been identified, for example, it is usually unnecessary for the customer to identify other firms merely because further consignments of the same kind need to be carried, or because the number of consignments is higher than anticipated.

Not all of the customer's consignments, however, may be the same. Cargoes can differ both in their stowage requirements and in the destinations to which they are sent. It is a great economy to the customer if all his shipping requirements can be met by the same firm, for then he only needs to make contact with one suitable firm. Such economy is, however, much easier to achieve in some cases than in others. A customer that despatches only a limited number of products to the same few destinations imposes fewer demands on the scope of the shipping firm than does a customer with a wide range of products sent to many different destinations. Thus a customer that has only a low level of diversification by product and by destination requires only a low level of diversification in the services offered by the shipping firm, while a highly diversified customer requires a high level of diversification in the shipping firm as well. In common, therefore, with other service industries - such as banking - where firms predominantly supply services to other firms, the minimisation of transaction costs encourages the supply firm to match its own diversification to that of its typical customer.

A very large customer may generate sufficient traffic to fully occupy vessels on certain routes, but in the liner trades this hardly ever occurs. The smaller the average amount of traffic generated by the typical customer, the more customers the shipping firm needs in order to keep its fleet fully utilised. The larger the number of customers, the more heterogeneity there is likely to be in terms of the cargoes they generate and the destinations to which they are sent. Hence, the smaller the 'size' of the typical customer, the greater is the need for the shipping firm to diversify if it is comprehensively to meet everyone's needs.

Another implication of fixed transaction costs is that the typical customer prefers one single firm to assume overall responsibility for the quality of service supplied. When a consignment is damaged or delayed in transit, the customer does not wish to have to sort out for himself the attribution of responsibility between the owners of the vessels used on the different legs of the journey, the stevedores concerned with loading, unloading and transhipment, the port authorities involved, and so on. It is much simpler if a single party agrees to indemnify the customer (for a fee) against poor service, however caused. The insurer then attempts to compensate himself for the damages paid over to the customer. If the insurer regularly deals with the various parties involved, and regularly handles complaints on behalf of many different customers, then his expertise, and his ability to sanction the culprits are both enhanced, and so the overall cost of recovering damages is reduced.

The provision of comprehensive insurance to customers poses particular problems, however. To begin with, the customer may not trust the insurer to pay up when required. The insurance contract may be so complex that the customer suspects there may be loopholes specially created by the insurer. This is an acute case of the general problem of 'asymmetric information' which arises wherever the supplier is the 'expert' and the customer the 'layman'. In this context, clarity and brevity in the specification of the service being supplied, and a reputation of 'fair dealing' by the supplier, are all important in creating trust. So too is a reputation for financial security, which means that the customer believes that the supplier is unlikely to renege on the contract because of financial exigency.

A further problem is that a firm which insures shipping services may be reluctant to insure against systematic errors which it cannot control. A major cause of such errors is bad management in a subsidiary activity, such as the control of traffic near ports. Customer demands for comprehensive insurance encourage the insurer to integrate backwards into these activities that are crucial for overall quality of service. [9] In the absence of institutional obstacles, this encourages the insurer to take a controlling interest, not only in the operation of ships along the routes on which he insures traffic, but in port operations as well.

7 Enhancing production qualities through a licensing system

The distinction between transactions qualities and production qualities is important in any industry, but it is particularly significant in the shipping industry. Transactions qualities are basically qualities that individual companies can control themselves through private investment in marketing activity, as explained above. Because of system effects, however, it is difficult for the individual company to control all the production qualities on the routes over which it operates. The overall quality of service will depend, for example, upon whether or not the sailing times of rival firms coincide with its own, or interleave with them. Unregulated competition could mean that several vessels belonging to different firms depart at the same time for the same port on the same route, which is clearly inefficient from the point of view of overall quality of service. It is worth noting, for future reference, that this problem reflects the element of 'natural monopoly' in the shipping industry, which arises from economies of scale in the design and operation of vessels. [10] Natural monopoly is a characteristic which shipping shares with other industries such as power generation and distribution, railways, broadcasting and telecommunications.

The importance of system effects indicates that, where production qualities are concerned, the competitive advantage of the European shipping line depends not only upon its own efforts, but upon the efforts of those who organise the competitive framework within which its services are supplied. This suggests that shipping policy may have an important role not merely in regulating shipping firms' access to markets, but providing an institutional framework which coordinates their activities efficiently.

To a limited extent, existing institutions already perform these functions. EEC policy condones closed conferences, and closed conferences have the potential to rationalise the operations of their members. In line with the remarks above, rationalised conferences tend to concentrate on delivering production qualities, and delegate the delivery of transaction qualities to individual member firms.

The question arises, however, as to whether the conference system is entirely adequate to deal with all of the system effects noted earlier. The existing system has had difficulties in coming to terms with competition between intermodal routes, and between the container consortia operating over each route. This reflects the general problem that most conferences are confined to a single route (or even a single direction of a single route) which is typically just one stage in a consignment's itinerary, and just one leg of a vessel's voyage.

Super-conferences are now emerging, whose activities have a wider scope than those of the traditional conference, and they may provide an opportunity to deal more comprehensively with system effects. Super-conferences are, however, potentially large organisations, and one of the dangers of large organisations is that they become bureaucratic and conservative in their outlook. Large organisations, whether private or state-controlled, are commonly alleged to have a poor record in innovation. Bureaucratic mismanagement is particularly difficult to detect in monopolistic organisations, since no yardstick based upon the performance of rival organisations is available to assess the efficiency of management.

In considering a suitable institutional framework for the shipping industry, therefore, the advantages of central organisation - in particular the ability to take a 'system view' when planning operations - must be traded off against the advantages of decentralisation in maintaining managerial accountability. In this context, the analogy between shipping and other naturally monopolistic industries, suggests that a system of regulated competition might be a suitable alternative to the present conference system. Regulation is, of course, more difficult to implement at the international level than at the national level, though if effected on a Community basis the problems might not be too acute, provided it were confined to routes dominated by European lines.

It could be objected that the scope for useful regulation of shipping is limited by existing and prospective cargo reservation agreements. Cargo reservation is, however, essentially, an entitlement, and does not commit the holder of the entitlement actually to implement it. This raises the possibility that the entitlement might be bargained away if the holder of the entitlement were offered suitable compensation. If regulation effected a genuine improvement in service, then access to an improved service for an export trade might be one element in such compensation, together, perhaps with concessionary rates on certain cargoes.

The form of regulation most suitable for the shipping industry appears to be a licensing or franchising system. The regulatory authority establishes a skeleton timetable of departures, designed to provide acceptable frequencies on each route, suitable connections between trunk routes and feeders, and as much through working as possible to minimise transhipment. The timetable is revised at fixed intervals in the light of the load factors experience on the various services.

The authority could be managed by a board consisting of representatives from the major shipping lines involved, the major customers, governments, and so on, though specific timetabling decisions would normally be devolved to professional managers who would be recruited on a non-political basis.

Individual shipping lines would bid competitively for licences. Ideally, competition should be completely open, and not restricted to operators from particular countries. The object of the exercise is to build up the competitive strength of European lines on the routes to and from Europe, and on the principal cross-trades on which they are involved, but this cannot easily be done if they are immune to outside competition. The object is to provide a competitive framework which stimulates the development of European managerial skills, rather than to offer protection which encourages management to relax.

Each licence would commit the licensee to operate the service to precise standards of time-keeping, quality of stowage, and so on, and repeated failure to comply would result in the black-listing of the offender. The licensee would have to agree not to make unlicensed sailings which compete with licensed services. The authority would need to invest quite heavily in the monitoring of licensees to check that these conditions are complied with.

Licences could be offered for either one-off sailing, for a sequence of sailings at prescribed intervals, or for all the sailings on a particular route. In the case of repeated sailings, the duration of the licence could be short term (say one year) or long term (5-10 years). The most suitable term of the licence will depend both on the reputation of the operator and the nature of the trade. It would be unwise to award a long term licence to a new entrant until he had demonstrated his proficiency in fulfilling a succession of short term licences. The nature of the trade is important because if special vessels have to be designed for a particular service then it is reasonable to confer the licence for a substantial length of time - although the firm whose licence expires before its fleet wears out always has the option of trying to sell its vessels secondhand to the firm that has outbid it when the licence was up for renewal. If, on the other hand, standardised vessels can be used, which are easily re-deployed to other services, then there is a good case for stimulating experimental entry by potential competitors by offering short term licences which are valid only for a few voyages at a time (the ability to switch vessels in and out of the service means that the market is essentially 'contestable' [11]).

Competition for licences will tend to select the most efficient operator because this operator, having lower costs, will be able to outbid its rivals for the licence. It also leaves individual firms free to bid for particular services so that they can roster their vessels in an efficient manner. Services whose licences cannot command a positive fee, and cannot be given away, will not operate unless someone is willing to supply a subsidy. The existence of a 'gap' in the scheduled services, however, is a signal to the authority that the skeleton timetable needs to be re-examined at the next opportunity.

To provide insurance for a bidder who is unsuccessful in securing a licence that is crucial to the efficient operation of his fleet, licences could be made transferable between operators (provided the operator acquiring the licence enjoys suitable status with the authority). This would create a resale market in licences, allowing firms that missed a particular auction to obtain the licences they require secondhand.

It would even be possible to tolerate a certain amount of intermediation by brokers and speculators in the market for licences. Brokerage in licences could reduce shipping firms' transaction costs in attending auctions, whilst speculative intermediation would provide a vehicle for shipping firms to hedge risks arising from an unforeseen need to add or drop sailings from fleet operations. Speculation may also help to smooth out fluctuations in the prices of licences by allowing professionals to anticipate market requirements. There may, however, be a case for restricting speculative holdings of licences to accredited individuals and institutions, to avoid destabilisation of the market.

The scope for intermediation will depend upon how far in advance of the sailings the licences are issued. Issuing licences well ahead of the departure time would allow firms to roster their vessels using a long term planning horizon, whilst access to a resale market would give them the freedom to rearrange these plans as required. On this basis, it would be reasonable to issue licences up to, say, two years ahead of the first sailing.

Because the skeleton timetable sets inviolable intervals between departure times, each operator has a small element of intertemporal monopoly on the services he operates. In some cases the degree of monopoly may be so small that the problem of monopolistic pricing is negligible. Predatory pricing, though, might still be a problem, in which, say, one firm threatens to spoil the market for a rival by cutting prices on earlier or later services if the rival bids a particular service away from the firm. In such cases, some form of price regulation may be called for. It is not essential that the timetabling authority also regulates the prices, although there is considerable logic in linking the two activities.

The introduction of a regulatory system of this kind seems to many people to be fraught with difficulties. Although licensing schemes have an eminent intellectual history they have rarely found favour with practical policy-makers. [12] It is worth noting, though, that the licensing of voyages was used by at least one Italian city-state in the fifteenth century, and that the concept of a resale market in licences is not too different from the concept of a voyage charter market as current operated. Licensing has been used in broadcasting and cable television, and in the commissioning of new railway lines. Licences for scheduled air services were issued by lottery in the United States quite recently.

Many of the practical difficulties anticipated with licensing reflect a fear that transactions qualities may suffer in the pursuit of enhanced production qualities. In the shipping industry, however, it is possible that regulation might actually improve transactions qualities. Some of the strategic reasons for the secrecy of conferences about their price would be eliminated, suspicions about monopoly conference profits would be partially allayed, and companies would therefore be freer to concentrate upon the aggressive marketing of the services they had elected to operate. Information supply by the regulatory authority would be available to confirm (or deny) claims made by operators about their quality of service. Although administrative difficulties would still be encountered, imaginative managers should not find these insuperable.

Appendix

This appendix presents the formal model that is the basis for the discussion of system effects on networks in section 5. The network is assumed to have a triangular configuration and to be unimodal. It is assumed that the origins, destinations and quantities of the various consignments to be shipped are pre-determined, and they remain fixed throughout the analysis. All the variables in the model are non-negative unless otherwise stated. The model generates, in principle, a trade-off between system cost and an index of system production quality. The model is not readily soluble, however. Programming techniques have a role in optimising certain aspects of system performance, but the only general method of deriving the precise form of cost-quality 'efficiency frontier' is by simulation. In view of the large number of variables that enter into the model, however, simulation is a very complicated exercise.

Demand for consignments

Let a_{ik} be the amount of the kth cargo consigned each period between two quays that, if routed directly, would be consigned along lane i (i=1,...,6; k=1,...,K). Let x_{ik} be the amount that is actually routed directly, and y_{ik} the amount that is routed indirectly on a through voyage. Let z_{ik} be the amount that is routed indirectly with transhipment at an intermediate point. Note that in the simple model considered here, no more than one transhipment is ever needed. From these definitions it follows that

(1) $$a_{ik} = x_{ik} + y_{ik} + z_{ik}$$

Let u_{ik} be the amount of the kth cargo actually transported along lane i. Then

(2) $$u_{ik} = x_{ik} + \sum_{m=1}^{6} \theta_{im} (y_{mk} + z_{mk})$$

where $\theta im = 1$ if the indirect alternative route to lane m involves the use of lane i, and is zero otherwise (for a tabulation of the θ-values, see Table 5.7).

Quay capacity:

It is assumed that each quay can handle the loading and unloading implied by the pattern of consignments a_{ik} (i=1, ...,6, k=1...,k) specified above. It can also handle transhipments when required, though at extra cost (see below). Congestion can, however, occur at the nodes, as described later.

Supply of voyages

Consider now the jth voyage (j=1,...,12). Let f_j be the frequency of sailings on the jth voyage, and let d_j be the capacity of the vessels used. Let h_{jk} be the proportion of cargo space devoted to the kth cargo. Let b_k - the bulk factor for the kth cargo - be the amount of space occupied by one unit of the kth cargo. Let e_{jk} be the maximum amount of the kth cargo that, within each period, can be carried along any leg of the jth voyage. Then

(3) $$e_{jk} = f_j d_j h_{jk} / b_k$$

Size of vessel

It is assumed that for reasons of technology, or because of limited dock-size, there is an upper limit to the size of vessel. Because of conditions at sea, there may be a lower limit on the size of vessel too. It is assumed that these limits are specific to each voyage. Let d_j^{min} be the minimum size of vessel that can operate on the jth route, and d_j^{max} the maximum size; then

(4) $$d_j^{min} \leq d_j \leq d_j^{max}$$

Overall carrying capacity by lane

Let $\bar{u}_{ik}$ be the maximum amount of the kth cargo that can be transported along the ith lane during one period.

$$(5) \qquad \bar{u}_{ik} = \sum_{j=1}^{12} \gamma_{ij} e_{jk}$$

where $\gamma_{ij} = 1$ if the jth voyage uses the ith lane, and is zero otherwise (for a tabulation of the γ-values, see Table 3).

The overall capacity constraint for the kth cargo on the ith lane is

$$(6) \qquad \bar{u}_{ik} \geq u_{ik}$$

where $\bar{u}_{ik}$ is given by (5), and u_{ik} by (2)

Capacity of indirect routes

Let $\bar{y}_{ik}$ be the maximum amount of the kth cargo that can be transported without transhipment by an alternative route to lane i. Then

$$(7) \qquad \bar{y}_{ik} = \sum_{j=1}^{12} \text{Ø}_{ij} e_{jk}$$

where $\text{Ø}_{ij} = 1$ if the jth voyage provides a through alternative to the ith lane, and is zero otherwise (for a tabulation of ø-values, see Table 8).

The indirect traffic capacity constraint for the kth cargo being routed around the ith lane is

$$(8) \qquad \bar{y}_{ik} \geq y_{ik}$$

Combining (2), (6) and (8), and using the non-negativity of the x,y and z variables, gives capacity constraints for direct consignments and transhipped consignments as well.

Distance travelled

Let μ_i be the distance from beginning to end of the ith lane. It is assumed that the distances at port between the node, the quay and the terminal are negligible. Let m_j be the total distance travelled each period by the fleet of vessels allocated to the jth voyage; then

(9) $$m_j = f_j \sum_{i=1}^{6} \gamma_{ij} \mu_i$$

Travelling time

It is assumed that all traffic travels along each lane at the same speed, s. Thus the travelling time associated with the jth voyage is

(10) $$t_{ij} = \sum_{i=1}^{6} \gamma_{ij} \mu_i s$$

Delays

Let g_h be the traffic density at the hth node: then

(11) $$g_h = \sum_{j=1}^{12} \psi_{hj} f_j$$

where ψ_{hj} is the number of times that the jth voyage arrives and departs from the hth node (h=1,2,3;j=1,...,12) (For a tabulation of ψ -factors, see Table 5.9).

Let $\bar{g}_h$ be the traffic capacity associated with the hth node. There is a time delay, t_{2h}, experienced by all voyages at the hth node, which is an increasing, and indeed accelerating, function of capacity utilisation at the node:

$$t_{2h} = t_{2h} (g_h / \bar{g}_h) \tag{12}$$

$$t'_{2h}, t''_{2h} > 0 \tag{13}$$

The total delays on the jth voyage are

$$t_{2j} = \sum_{h=1}^{3} \psi_{hj} t_{2h} \tag{14}$$

Stowage time

Let t_{3j} be the total stowage time associated with the jth voyage. It is a function of the number of ports of call and the stowage problems posed by the itinerary. It is, however, assumed to be independent of the quantity of cargo to be stowed. The stowage problems are set out in Table 5.4. The values of the s-elements are set out in the final column of Table 5.9.

Voyage time

The total voyage time, v_j, is the sum of the travelling times, the delays and the stowage times:

$$v_j = \sum_{m=1}^{3} t_{mj} \tag{15}$$

Number of vessels required

Vessels are laid up for maintenance at the terminals where they are based. It is assumed that the time allotted for maintenance is a constant fraction ζ_j of the total voyage time v_j.

Let n_j be the number of vessels required to maintain services on the jth voyage. Then

$$(16) \qquad n_j = f_j v_j (1 + \zeta_j)$$

Cost structure

Four components of shipping cost may be distinguished: capital costs, time-related operating costs, distance-related operating costs, and transhipment costs. Costs which are specifically speed-related are ignored. The costs of loading and unloading consignments at the beginning and end of their journeys are ignored since these are determined uniquely once the pattern of consignments has been specified. These costs do not, therefore, affect the choice of an efficient organisation of shipping services.

Consider first the purchase price of a vessel. It is assumed that the construction cost, and hence the purchase price, is related to its size in a manner which reveals increasing returns to scale. Thus the price, p_j, of a vessel operating on the jth voyage is given by

$$(17) \qquad p_j = \alpha_{10} + \alpha_{11} d_j$$

where $\alpha_{10}, \alpha_{11} > 0$ are fixed parameters and d_j is the capacity of the vessel. Allowances for interest charges, depreciation, inflation - adjustment of capital values, and taxation are reflected in the coefficient r_{1j}, which is applied to the purchase price to calculate the capital charges per period. The capital cost of the fleet allocated to the jth voyage is then

(18) $$c_{1j} = r_{1j} (\alpha_{10} + \alpha_{11} d_j) n_j$$

The time-related operating costs include the wages of the crew, stores, etc. It is assumed that these costs too reveal economies of scale to the size of the vessel. Let r_{2j} be the input price for time-related operating services; then time-related operating costs for the fleet on the jth voyage is

(19) $$c_{2j} = r_{2j} (\alpha_{20} + \alpha_{21} d_j) n_j$$

where $\alpha_{20}, \alpha_{21} > 0$ are fixed parameters.

Let r_{3j} be the price of distance-related operating services. It is assumed that the input of these services - which include maintenance of engines etc. - is directly proportional to the distance travelled and - like the other components of cost - is independent of how fully the vessel is loaded. It is assumed that the cost is an accelerating function of the speed at which the vessel travels. Thus the distance-related operating costs for the fleet on the jth voyage are

(20) $$c_{3j} = r_{3j} (\alpha_{30} + \alpha_{31} d_j) w (s) m_j$$

where $\alpha_{30}, \alpha_{31} > 0$ are fixed parameters and

(21) $$w' , w'' > 0$$

It is assumed that the input of services required to tranship a cargo is directly proportional to the amount of cargo involved. It is also assumed that the price of transhipment services, r_4 is the same at all quays. Thus the total cost of transhipment is

(22)
$$c_4 = r_4 \sum_{k=1}^{K} \alpha_{4k} \sum_{i=1}^{6} z_{ik}$$

where $\alpha_{4k} > 0$ are fixed parameters (k=1,...,K).

Total shipping cost, c, is

(23)
$$c = \sum_{j=1}^{12} \sum_{m=1}^{3} c_{mj} + c_4$$

In the special case where $\alpha_{m0} = \alpha_0, \alpha_{m1} = \alpha_1$ (m=1,2,3) and $\alpha_{4k} = \alpha_4$ (k=1,...,K) this takes on the simple form

(24)
$$c = \sum_{j=1}^{12} (\alpha_0 + \alpha_1 d_j)((r_{1j} + r_{2j}) n_j + r_{3j} wm_j) + r_4 \sum_{k=1}^{K} \sum_{i=1}^{6} z_{ik}$$

Note that because the prices r_{mj} are j-dependent they allow, in principle, for the fact that the prices of services may vary depending upon the node at which the voyage starts and terminates (j=1,2,3).

Production quality

The final step in specifying the model is to construct an index of production quality. The index is an aggregate of indices constructed for each type of cargo. For each type of cargo, the quality index is a weighted sum of seven different aspects of quality, the first two of which enter positively, and the remainder negatively, into the index.

The seven aspects are

(1) frequency of service;

(2) amount of spare capacity available;

(3) the amount of indirect routing of cargoes;

(4) the amount of transhipment of cargoes;

(5) normal voyage time;

(6) time delays to voyages; and

(7) amount of stowage time.

Each aspect is given two weights. The first w_{mk}, indicates the importance of the mth aspect to the kth cargo. The second is an approximate measure of the quantity of cargo affected. The measure is approximate only because the calculation of the true quantity actually affected by each aspect of quality is quite complicated. It simplifies the algebraic formulae considerably, therefore, to use a suitable approximation. Thus to measure the impact of the first aspect - frequency of service - the frequency of departures is calculated for each lane, and then weighted by the amount of the kth cargo carried over that lane; the weighted frequencies are then aggregated over the lanes, and a further weighting then applied for each cargo, w_{1k}, to determine an index of 'frequency quality' for the kth cargo. The overall 'frequency quality' is then measured as the sum of the qualities for each of the cargoes.

The application of similar approximations for the other aspects of quality gives the intuitively reasonable index of over quality,

$$q = \sum_{k=1}^{k} w_{1k} \sum_{i=1}^{6} \left(\sum_{j=1}^{12} \gamma_{ij} f_j \right) u_{ik}$$

$$+ \sum_{k=1}^{k} w_{2k} \sum_{i=1}^{6} u_{ik}$$

$$- \sum_{k=1}^{k} w_{3k} \sum_{i=1}^{6} y_{ik}$$

$$- \sum_{k=1}^{k} w_{4k} \sum_{i=1}^{6} z_{ik}$$

$$- \sum_{k=1}^{k} w_{5k} \sum_{i=1}^{6} \mu_i u_{ik}$$

$$- \sum_{k=1}^{k} w_{6k} \sum_{h=1}^{3} \left(\sum_{j=1}^{6} \psi_{hj} e_{jk} \right) t_{2h}$$

$$- \sum_{k=1}^{k} w_{7k} \sum_{j=1}^{12} e_{jk} t_{3j}$$

Figure 1

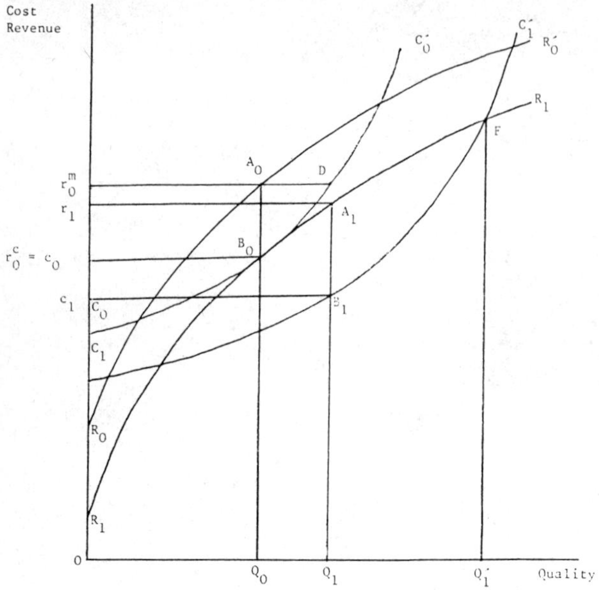

Figure 2

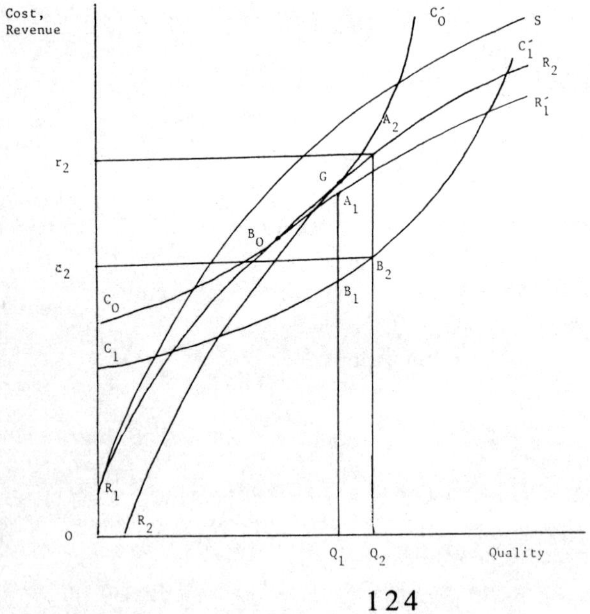

Figure 3

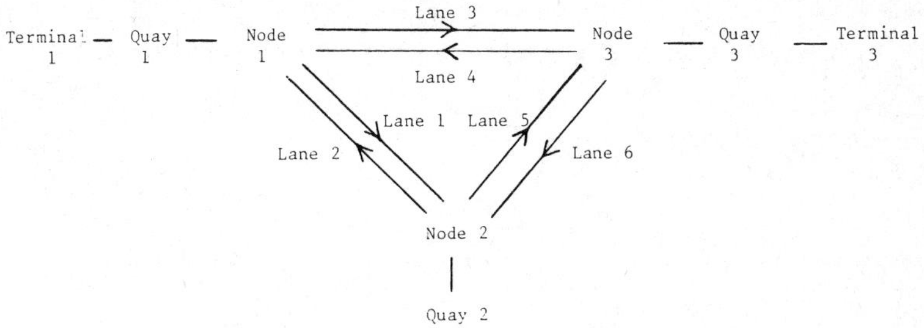

Note: The discussion in this paper assumes that vessels always call at the the quay associated with a node they pass through. There are no 'express' services which pass through a node without stopping at the quay.

Figure 4

Specific interpretations of the network

(a) Atlantic-version I

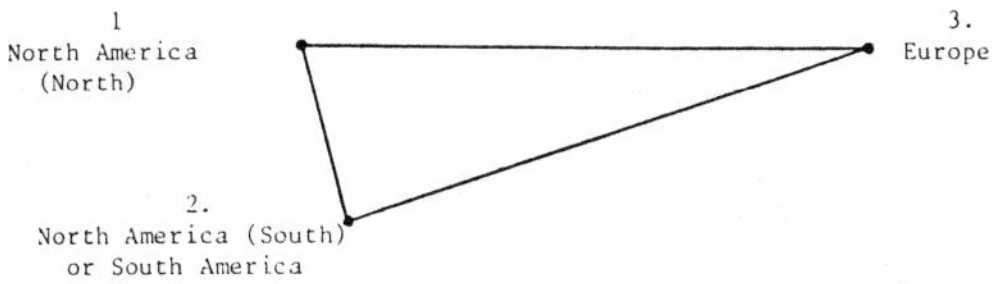

(b) Atlantic-version II

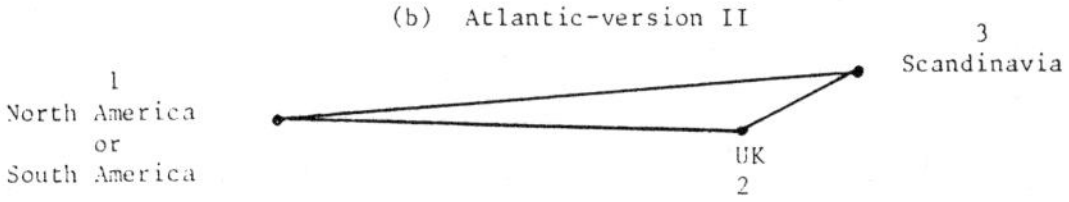

(c) World

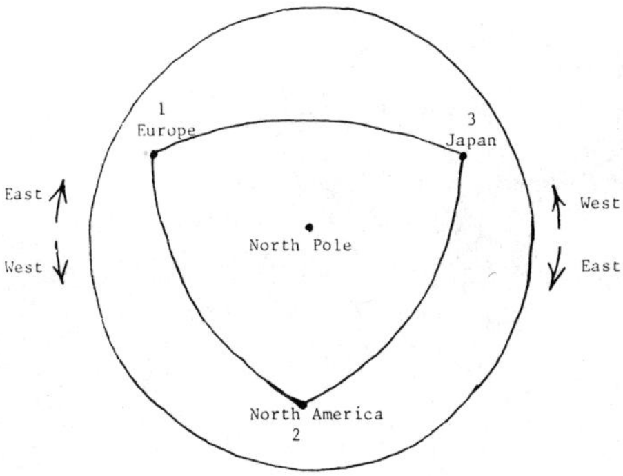

Figure 5

Specific interpretations of a restricted network

(a) Multi-modal network

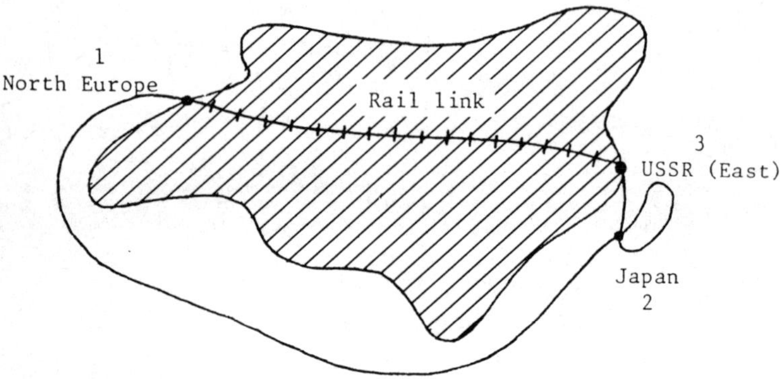

(b) Trunk route with feeder

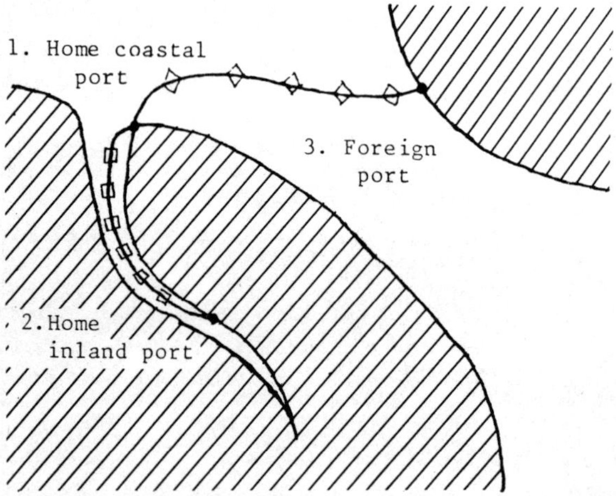

Table 5.1: Production Qualities

A Incidence of damage

1 How well equipped is the vessel?
2 How many different kinds of stowage are offered?
3* What proportion of consignments are transhipped? (as C5)

B Spare capacity

1 What is the risk that last minute consignments are refused for lack of space?

C Duration of transit

1* What are the speeds along the various lanes?
2 What are the times allotted for loading and unloading? How easy are stowage arrangements?
3* What are the delays caused by congestion at the ports?
4* What proportion of consignments are routed indirectly?
5* What proportion of consignments are transhipped? (As A3)
6 What is the average time waiting for a connection during transhipment?

D Frequency of service

1* What are the frequencies of the different voyages?
2 Is the service on a given voyage provided by inter-leaved sailings of specialised vessels, or by a complete fleet of multi-purpose vessels?

E Uncertainty

1 What is the risk that last minute consignments are refused for lack of space (as B1)
2* What are the delays caused by congestion at the ports (as C3)
3* What proportion of consignments are transhipped? (as C5 and A3)

Note: * indicates that the question is considered in the model presented in the appendix

Table 5.2: Transactions Qualities

A Clarity of specification of the service

1 How precisely is the nature of the stowage offered defined, and its suitability for particular types of cargo explained?

B Accessibility of the supplier

1 How easy is it to make contact with the shipping service? How well is it advertised?

2 How comprehensive is the service? In other words, is it possible for the customer to use a single contact for all his traffic?

C Ease of negotiation

1 How easy is it to fix a price? For example, does the shipping service have a freely quoted price?

2 Is it possible for the shipper to choose the terms of the contract so that he does not have to pay for 'extras' he does not want, or go without services he deems essential, because they are not in popular demand.

3 Can reservations of space be made well in advance if necessary, and cancelled with reasonable notice when not required?

D Comprehensive insurance

1 Is comprehensive insurance available against costs arising from delays and damage?

E Reliability of contractual enforcement

1 Is there a single locus of responsibility for the customer to claim from in the event of a failure in the shipping service?

2 Does the bearer of responsibility have a good reputation, and the financial backing to maintain it?

Table 5:3: Return Voyages with up to four Legs

Voyage index j	Description	Itinerary					Lanes used (index i) 1	2	3	4	5	6
From Terminal 1												
1	Shuttle to quay 2	1	2	1			1	1	0	0	0	0
2	Shuttle to quay 3	1	3	1			0	0	1	1	0	0
3	Clockwise round trip	1	3	2	1		0	1	1	0	0	1
4	Anti-clockwise round trip	1	2	3	1		1	0	0	1	1	0
5	Return loop to quay 2	1	3	2	3	1	0	0	1	1	1	1
6	Return loop to quay 3	1	2	3	2	1	1	1	0	0	1	1
From Terminal 3												
7	Shuttle to quay 2	3	2	3			0	0	0	0	1	1
8	Shuttle to quay 1	3	1	3			0	0	1	1	0	0
9	Anti-clockwise round trip	3	1	2	3		1	0	0	1	1	0
10	Clockwise round trip	3	2	1	3		0	1	1	0	0	1
11	Return loop to quay 2	3	1	2	1	3	1	1	1	1	0	0
12	Return loop to quay 1	3	2	1	2	3	1	1	0	0	1	1

Notes: The itineraries from terminal 3 may be derived from the itineraries from terminal 1 by interchanging the numbers 1 and 3 in the itinerary sequence.

The transpose of the 12 x 6 matrix of elements on the right hand side of the table is equivalent to the matrix T.

Table 5.4: Stowage Properties of Voyages from Terminal 1

Index j	Overall ease of stowage	Stowage Procedures
1	Easy	Stow in any order at both quays 1 and 2
2		
3	Difficult	Stow at quay 1 with goods for quay 2 stowed below those for quay 3. At quay 3, goods for quay 1 need to be placed below those already stowed at quay 1 for unloading at quay 2.
4		Stow at quay 1 with goods for quay 3 stowed below those for quay 2. At quay 2, goods for quay 1 need to be placed below those already stowed at quay 1 for unloading at quay 3.
5	Moderate	Stow at quay 1 with goods for quay 2 stowed below those for quay 3. Goods for quay 2 only are loaded at first call at quay 3; they can be loaded in any order. At quay 2 the vessel is emptied and goods for quay 1 are loaded below those for quay 3. At the second call at quay 3, only goods for quay 1 are loaded, and they can be stowed in any order.
6		Stow at quay 1 with goods for quay 3 stowed below those for quay 2. Goods for quay 3 only are loaded at first call at quay 2; they can be loaded in any order. At quay 3 the vessel is emptied and goods for quay 1 are loaded below those for quay 2. At the second call at quay 2, only goods for quay 1 are loaded, and they can be stowed in any order.

Table 5.5: Ease of Stowage for Voyages from Terminal 3

Index, j	Overall ease of stowage
7	Easy
8	
9	Difficult
10	
11	Moderate
12	

Table 5:6 Direct and Indirect Routing of Consignments

Consignment From	To	Direct route *via* lane	Indirect route *via* lanes
1	2	1	3 , 6
2	1	2	5 , 4
1	3	3	1 , 5
3	1	4	6 , 2
2	3	5	2 , 3
3	2	6	4 , 1

Table 5.7: Correspondence between lane use on direct routes and lane use on indirect routes

Lane used on direct route (index i)	Lanes used on indirect route (index i)					
	1	2	3	4	5	6
1	0	0	1	0	0	1
2	0	0	0	1	1	0
3	1	0	0	0	1	0
4	0	1	0	0	0	1
5	0	1	1	0	0	0
6	1	0	0	1	0	0

Note: The 6 x 6 matrix of elements in this table is equivalent to the matrix Θ.

Table 5.8: Availability of through voyages for indirect routings

Voyage (index j)	Indirect alternative route to lane (index i) 1	2	3	4	5	6
1	0	0	0	0	0	0
2	0	0	0	0	0	0
3	1	0	0	1	0	0
4	0	1	1	0	0	0
5	1	1	0	0	0	0
6	0	0	1	1	0	0
7	0	0	0	0	0	0
8	0	0	0	0	0	0
9	0	0	1	0	0	1
10	0	0	0	1	1	0
11	0	0	0	0	1	1
12	0	0	1	1	0	0

Note: The transpose of the 12 x 6 matrix of elements in this table is equivalent to the matrix Φ.

Table 5.9: Traffic Densities Through Nodes, and Stowage Times, by Voyage

Voyage (index j)	Nodes (index h) 1	2	3	Stowage time factor per node	Stowage time per voyage
1	1	1	0	σ_1	$2\sigma_1$
2	1	0	1	σ_1	$2\sigma_1$
3	1	1	1	σ_3	$3\sigma_3$
4	1	1	1	σ_3	$3\sigma_3$
5	1	1	2	σ_2	$4\sigma_2$
6	1	2	1	σ_2	$4\sigma_2$
7	0	1	1	σ_1	$2\sigma_1$
8	1	0	1	σ_1	$2\sigma_1$
9	1	1	1	σ_3	$3\sigma_3$
10	1	1	1	σ_3	$3\sigma_3$
11	2	1	1	σ_2	$4\sigma_2$
12	1	2	1	σ_2	$4\sigma_2$

Note: The transpose of the 12x3 matrix of elements on the left hand side of the table is equivalent to the matrix ψ

It is assumed that $\sigma_1 < \sigma_2 < \sigma_3$

The right hand column of the table is equivalent to the vector t_3.

NOTES

I am grateful to Bernard Gardner for extensive discussions on an earlier draft of this chapter. I should also like to thank the discussants at the Conference on the International Maritime Policy of the European Community for their comments. Thanks are also due to George Yannopoulos for his encouragement during the preparation and revision of this chapter, and to my colleague Michael Utton for technical advice.

1 See Schrier, Nadel and Rifas (1985) and Tzoannos (1986).

2 There are two main reasons why this argument appears plausible. First there is an enormous number of ways in which governments can subsidise their national shipping lines, and many of the indirect methods are easy to conceal. Methods include special tax treatment of shipping firms, or more generally, the use of a tax regime that benefits capital-intensive industries; privileged access to subsidised shipbuilding; poor safety standards; low living standards for the crew, and the conscription of merchant seamen; offers of lucrative government cargoes, providing revenue to cross-subsidise rates on other cargoes; subsidised fuel bunkers; and so on. Secondly, there is the argument that because of the international nature of shipping operations, international differences in the true opportunity costs of resources do not provide the same basis for national competitive advantage in the shipping industry that they do in manufacturing industries. In the absence of subsidies and other selective measures all operators on a given route would have the same access to resources at each point along the route, and so would have the freedom to choose the cheapest resources, irrespective of their nationality. In the shipping industry, therefore, a national operator on a route cannot benefit from cheap indigenous resources unless other operators on the same route are denied access to these resources on similar terms through discriminatory intervention.

3 The comparison suggested here is with the conventional theory of monopoly as articulated in, for example, Chamberlin (1933).

4 A review of some of this evidence is given in the introduction to Casson (1986b).

5 See, for example, Casson and associates (1986), Chapter 1.

6 System effects arise in most transport industries. For a review of recent economic models of transport systems see Daughety (1985).

7 For a comprehensive survey of the properties of vehicle flows over networks, and the optimisation problems associated with them, see Bodin, Golden, Assad and Ball (1983).

8 On the general theory of transaction costs see Williamson (1975) and Casson (1982).

9 Other motives for backward integration in the shipping industry are discussed in Casson (1986a).

10 On the definition and application of the concept of natural monopoly see, for example, Sharkey (1982).

11 Contestability is discussed by Baumol, Panzar and Willig (1982).

12 For the recent review of franchising issues, and an application to the broadcasting industry, see Domberger and Middleton (1985). The seminal reference is Demsetz (1968); see also Williamson (1976).

REFERENCES

Baumol, W J, J C Panzar and R D Willig (1982) *Contestable Markets and the Theory of Industry Structure* , New York: Harcourt Brace Jovanovich.

Bodin L, B Golden, A Assad and M Ball (1983) Routing and Scheduling of Vehicles and Crews: The State of the Art, *Computers and Operations Research, 10(2), pp 63-211.*

Casson, M C (1982) *The Entrepreneur: An Economic Theory*, Oxford: Blackwell.

Casson, M C (1986a) Vertical Integration in the Shipping Industry, *Journal of Transport Economics and Policy*, 20, pp 3-29.

Casson, M C (1986b) Quality Control and Vertical Integration, University of Reading Discussion Papers in Economics, republished in M C Casson, *The Firm and the Market: Studies in Transaction Costs and the Strategy of the Firm*, Oxford, Blackwell, 1987, Chapter 4.

Casson, M C and associates (1986) *Multinationals and World Trade: Vertical Integration and the Division of Labour in World Industries*, London: George Allen and Unwin.

Chamberlin, E H (1933) *The Theory of Monopolistic Competition : A Re-orientation of the Theory of Value*, Cambridge, Mass: Harvard University Press.

Daughety, A F (ed) (1985) *Analytical Studies in Transport Economics*, Cambridge: Cambridge University Press.

Demsetz, H (1968) Why Regulate Utilities? *Journal of Law and Economics*, 11, pp 55-66.

Domberger, S and J Middleton (1985) Franchising in Practice: The Case of Independent Television in the UK, *Fiscal Studies*, 6(1), pp 17-32.

Schrier E, E Nadel and B Rifas (1985) *Outlook for the Liberalisation of Maritime Transport*, London: Trade Policy Research Centre, Thames Essay No. 44.

Sharkey, W W (1982) *The Theory of Natural Monopoly*, Cambridge: Cambridge University Press.

Tzoannos, J (1986) The EEC Common Maritime Transport Policy and the Liberalisation of World Shipping Markets. This volume.

Williamson, O E (1975), *Markets and Hierarchies: Analysis and Antitrust Implications*, New York: Free Press.

Williamson, O E (1976) Franchise Bidding for Natural Monopolies - In General and with respect to CATV, *Bell Journal of Economics*, 7, pp 73-104.

CHAPTER 6

UNITED STATES AND EUROPEAN COMMUNITY INTERNATIONAL MARITIME POLICIES

Gerald Seifert

1 Introduction

This discussion is intended to compare the developing, common transport policy for the European Community with the established set of standards existing in the United States. Of equal importance with the comparison is the attempt to develop a basis for determining whether the developing rules will be harmonious with the regulatory scheme in the United States, and where conflicts may exist.

As with the Shipping Act of 1916 in the United States, the central concern of the maritime transport rules is to provide for exemptions to antitrust prohibitions similar to those the EEC Commission does for other agreements. Parenthetically, it is important to note that the need to develop these rules did not originate with a legislative determination, but rather a judicial interpretation making the Treaty of Rome, Title IV, applicable to sea transport, even though that title dealt only with rail, road, and inland waterway transport. In this latter respect, the first comparison is the analogous response by way of legislation to judicial activism - in the case of the EEC Commission the need to adapt Regulation 17 to the requirements of the maritime industry, and in the case of the United States Congress, the need to adapt the 1916 Act to the evolving expansion of the antitrust laws, the result being the Shipping Act of 1984.

At the outset, it may be illuminating to examine the differing premises in the approaches taken by the Congress of the United States and the EEC Commission. It would seem, as a matter of legislative policy that the Commission does not contemplate the kind of judicial erosion of legislative intent that was observed in the United States, where the courts, in the 60 year period following enactment of the 1916 Act, whittled away at the legislative mandate to such a degree that individuals subject to the Act found themselves constantly under threat of prosecution under the criminal and civil provisions of the Antitrust Laws. The basis for this judicial activity was the necessity for applicants to demonstrate that their actions were *in the public interest* and that they were, consequently, entitled to the *exemption* provided for in the 1916 Shipping Act. This appears to be the style employed by the Commission; that is, that there is primacy in the rules of competition, and that a conscious act of exemption must be effected in order to allow the type of concerted activity that is viewed as necessary if transportation services are to be available to the broadest population of users.

2 The 1984 Shipping Act

The drafters of the Shipping Act of 1984, obviously benefitting from the experience of over sixty years, employed a different tack, and *disapplied* the antitrust laws to an activity described in the Shipping Act resulting in the rules of competition governing ocean shipping in the United States being co-equal with the rules of competition embodies in the Antitrust Laws. Rather than being subsidiary to legislation prohibiting certain anti-competitive conduct, as is the style with the EEC common rules, the United States avoids ranking, by creating an independent set of rules governing the conduct of international, maritime, shipping practices.

The basis for granting the exemption in the EEC is, among others, the proviso that the shipping firms involved, charge nondiscriminatory rates. In addition, carriers would have to consult with transport users, to respect certain standards for loyalty agreements, to leave users the choice of associated services not covered by freight charges, to publish detailed tariffs, and to notify the Commission of the results of arbitration and conciliation proceedings.

There would be no provision for filing of conference tariffs. Instead, the Commission could act on complaints and investigate on its own initiative. These procedures parallel those of Regulation 17, and in some cases the proposal precisely echoes Regulation 17. Likewise, final procedural provisions for fines, periodic penalty payments, judicial review, hearings, professional secrecy and publication of decisions are virtually identical.

The adoption of the common rules for maritime transport shifts the balance of power in maritime matters from the individual states to the EEC Commission. Confirmation of the existence of abusive practices or of special circumstances (including restrictive measures by foreign governments) would withdraw the blanket exemption from the conference in question and simultaneously rule whether and in what conditions an individual exemption might be granted. Without an exemption the shipping firms would be unable to operate; their contracts with shippers would be void and they would be subject to civil suit from any shipping firm which had been excluded from the trade.

Enforcement of Commission decisions could include the imposition of fines and penalties of up to ten percent of the annual operating revenues of the offending firm. Commission action in response to actions of a foreign government requires the Commission to enter directly into consultations with the competent foreign authorities, which with subsequent Council authorisation could begin negotiations for an arrangement to resolve the situation.

Under the Shipping Act of 1984, the foregoing regulatory activity would be treated differently, mainly because of the difference in basic approach. Taken in the order as close as possible to the way presented above, the Shipping Act regulates as follows:

The granting of non-applicability of the antitrust laws is not contingent upon anything except the fact that the carriers are engaged in the type of activity listed as permitted for ocean common carriers and others subject to the act. Failure to discuss certain practices with users might, in certain cases, be found to be a prohibited act, but would not result in loss of antitrust immunity; unlike the common rules, the United States does not grant status to shippers' councils, and negotiations between conferences and councils is not contemplated as regulatory feature.

Loyalty agreements are outlawed in the United States, unless specific two-part agreements are sanctioned by the Department of Justice (loyalty agreements are outside the scope of the Shipping Act). As with the common rules, tariffs must be published, but arbitration and conciliation is not provided for.

Central to the regulatory scheme in the United States is the mandatory filing of tariffs and the concomitant enforcement, by the Federal Maritime Commission of those filed tariffs. Since rebating is not permitted in the United States, the basis for determining the violation of this prohibition is the existence of a baseline from which to measure the unlawful cuts - tariff filing provides this baseline. Likewise, the filing of tariffs is viewed as necessary, in the United States, to allow the Federal Maritime Commission to enforce the Controlled Carrier Law, that distinguishes the predatory pricing practices of state controlled carriers from market force practices by arraying the controlled carriers filed tariffs against those of the conference carriers with whom they compete.

Violation of specified prohibited acts by persons subject to the Shipping Act gives rise not to an elimination of the exemption from the Antitrust Laws (since the latter are disapplied) but rather to a set of civil penalties imposed by the Federal Maritime Commission. In the event of severe violations, the existence of filed tariffs permits the imposition of effective remedial action by suspending or revoking the tariffs of the wrongdoer, effectively suspending their operations, pending the elimination of the abuse. Private rights of action are not provided for since the entire scheme is based upon a total *public service* concept in which violations of the trust granted by the public are punished by the imposition of public rather than private sanctions. Injunctive relief may be sought (and granted) in either scheme.

3 Similarities and contrasts

Some of the elements of the common transport policy are not really capable of critical, comparative analysis, but should be addressed. First, the provisions dealing with cabotage (excluded from the compromise) are interlaced with political rather than legal conflict. The political debate in Europe is similar to the difficulties that would present themselves if proposals to phase out or eliminate the US Jones Act were presented to the Americal Congress. Cabotage is a traditional national privilege, and the mass of the United States itself, with its many coastlines, and its concern with national security militates against the type of proposal being considered by the Community.

The phase-out of bilateral agreements and prior consultation on matters that might affect the freedom of EEC shipping companies to provide services, rings a very familiar bell. The United States as a matter of general policy does not favour bilateral or multilateral shipping arrangements, especially if the underlying or principal purpose is to allocate or reserve cargo. One has only to compare the absolute resistance by the United States to the UNCTAD Code of Liner Conduct with the approach taken by the Europeans who would phase-out bilaterals. Traditionally, the United States has engaged in what migh be termed *responsive bilateralism*. Where other countries have imposed schemes to restrict access to cargoes in the foreign commerce of the United States, we have gone to the table to pare down the reservations traditionally taken by those other nations. There is a world of difference between the approach taken in the United States and that in which nations, as an extension of their former colonial mercantile systems have entered into trading agreements. Obviously, there is certainly prospective joy in the United States in consideration of the possibility that bilateral agreements in the Community may indeed by phased out, so that fair competition may determine who carries the cargo.

The designation of certain cargoes as reserved to certain national-flag carriers, and the proposal to eliminate such reservations is laudable and certainly consistent with the philosophy of fair trade that drives the United States. Unless cargoes are truly government cargoes, and by that is meant cargoes belonging to the government and being shipped for a government and not a commercial purpose, there is no basis for allocating or reserving the cargo to specified national flag carriers.

On unfair pricing practices by state-trading countries, the EEC approach is comparable to the US Controlled Carrier Act, and perhaps, may go even a bit further. The EEC approach includes the imposition of a redressive duty that is not available to the Federal Maritime Commission, and may indeed improve the enforceability of the EEC rules.

While the comparative analysis may not be exhaustive, it sets out the basis for determining whether harmony across the Atlantic will be encouraged or discouraged. The adoption of the four Regulations on 22nd December 1986 brings the US and the EEC closer, since competition and pro-competitive regulatory schemes have been the engine that has long driven maritime services in the US foreign commerce.

CHAPTER 7

EASTERN-BLOC SHIPPING POLICIES AND THE EUROPEAN SHIPPING INDUSTRIES

Patrick Gautrat

1 Introduction

In France, ever since Colbert, we have been convinced that shipping is highly specific in character. Throughout history, there have been few periods - even in Great Britain - when states have disregarded this activity which is commercial, but is also wholly international and closely connected with the geo-strategic imperatives of each country. Thus, national shipping strategies fare very badly with "stop-go" policies, and their projections are for the long term, as recalled a century ago now by the French statesman, Adolphe Thiers: "whoever speaks of a shipping fleet speaks of continuity, time and determination. The one thing among all that can least do without determination on the part of government, and ordered thinking, is shipping".

Even today, on various continents, we can see very well-drawn strategies at work for the development or maintenance of merchant fleets. However, without ignoring those countries' imperatives of sovereignty or objectives of power (particularly in the case of the United States), they are above all pursuing commercial ambitions. Our great mistake when we come to deal with shipping in State trading countries, is that, as it happens, we forget the "State" aspect, and we reason chiefly from the standpoint of trade.

The European Economic Community cannot imitate the shipping strategies of Taiwan and the USSR, nor the management style of Yang Ming or Morflot.

Trading strategy in the USSR is an instrument of super-power strategy and it must be regarded as being organised on military lines. The diplomat by training that I happen to be does not recoil from dealing with matters from their economic standpoint. However, in the matter in hand, one can only follow the Soviet Union Communist Party. While the First Secretary called for an improvement in the productivity of Soviet industry, he never referred to shipping in a speech that was, even so, very long. What better benchmark is there for judging that the fleet is a sector whose political dimension is such that it cannot be governed only by the criteria that apply to the remainder of the economy?

And then again, why mention only the USSR? Probably for the sake of simplicity, but perhaps also because Soviet tonnage represents two-thirds of all the COMECON fleets together, which in 1984 amounted to 30 million GRT. Without neglecting the importance of fleets such as those of Poland or East Germany, it is better to consider them within the overall context of work-sharing as practised in the COMECON. In the early 1960s, an *Independent Freight Coordination Office* was founded in Moscow, with the purpose of sharing the markets among the operators. Thus, as the years went by, the Polish Ocean Lines (POL) were observed to be setting themselves up on the Pacific and Australian traffic routes, and the East German DSR line setting up in Central America, while the Soviet Balt-Orient concentrated on the Far East. Better still, when they are in the same traffic, the companies allocate the roles: in Central America, Balt-Capas (USSR) negotiated with the Conference, while DSR remained an outsider. In the Far East, POL are in the FEFC Conference, while Balt-Orient is an outsider.

The Moscow *Bureau* is thus perfectly aware of the strategies of the major conferences, and this enables it to 'control the game'. In the final analysis, the sister-country fleets, whatever their own commercial ambitions, are chiefly the Soviet partner's 'top-up' resources, or they occupy positions in traffic where the Soviets do not wish to announce themselves overtly.

Therefore, in order to see exactly what is at stake for Europe in the development of the merchant fleets of the state trading countries, the analysis of the phenomenon should not be confined to just the commercial aspect. In part 2 of this chapter, the long-term strategy that explains this development is presented.

In part 3, which deals with the EEC's responses, the ideas advanced are obviously valid for the COMECON countries as a whole, but they apply first and foremost to the USSR.

2 A development based on a long-term strategy

At first sight, the Soviet fleet does not occupy a leading role on the world stage. It ranks only fifth by flag (with 16.8 million GRT, equivalent to 5% of the total). It should be noted, however, that for general cargo, i.e. liner traffic, it ranks first, with some 8 million GRT.

This predominance is explained by a deliberate strategic and historic choice, and this we will examine before considering the changes in the Soviet fleet, its structures and its market positions.

i Doctrine

The Second World War was won by the power which had mastery of the sea in military terms, but which was also capable of building more 'Liberty ships' than the enemy could destroy. The point was not lost upon Admiral Gorshkov, who was the Head of the Soviet Navy for thirty years, a devotee of the American Mahan, and who in 1976 was the author of a monumental book on 'The sea power of the State'. His central argument can be summarised as follows: 'Among the factors which characterise our country's military power, its sea power is taking an increasingly significant role. It is in that sea power that is expressed the true scope for actively using the oceans of the world to build Communism'.

If the USSR wishes to catch up with and overtake the United States, it must - despite its geographical handicaps - become the leading sea power. It is perhaps the mistake of many Western specialists to have ignored Russian seafaring tradition for too long and to have viewed the USSR as confined to the role of a purely continental power. While the Soviet Navy is continuing to expand, even after Gorshkov was ousted to be replaced by Admiral Chernavin, all the fleets under the Soviet flag have in the last twenty years undergone spectacular growth:

- the *fishing* and fish-processing fleet has 3,500 units (7.5 million GRT), equivalent to half the world tonnage. Its productivity is low, but its strategic and espionage role is fundamental.

- the *fleet of scientific research vessels* is also the world leader with 183 vessels totalling 320,000 GRT, equivalent to 47% of the world fleet. Half this fleet operates outside Soviet territorial waters.

- the *merchant fleet* is organised on quasi-military lines, with 17 companies organised into 3 fleets, the North-West, the Southern and the Far-East under joint management by Morflot. The latter controls not only the ships, but also the ports, repair yards and all the enterprises involved in the operation and maintenance of and assistance to the fleet.

Everyone knows what are the major motivations of this merchant fleet:

- *logistic support* to the fighting fleet, the effectiveness of which can be gauged either from its refuelling or repairs function (when the merchant fleet flies the fleet auxiliary flag) or from its amphibious action, as witnessed with the Soviet ro-ro ships in Ethiopia, when a few years ago, large quantities of military equipment had to be brought in as aid to the government on the spot. The container-carriers and ro-ro's are therefore the most useful vessels for the latter type of assignment.

- *a source of foreign currency* , and this is a motivation that can be found in all the COMECON countries, implying that positions are taken in the most lucrative markets, i.e. liner traffic of the OECD countries.

- *a commercial presence* , leading to a *political presence* in the countries of call. In this respect, the regular line affords the best opportunities for penetrating a foreign country, particularly a developed country, by the setting up of its commercial networks whose ambitions and tasks often exceed the bounds of the commercial domain alone.

Surely, then, it is not surprising that the main effort was focused on the liner fleet and not on bulk shipping.

ii The history of the Soviet fleet's development

Hans Jacob Kruse, the President of Hapag Lloyd, distinguishes four stages in recent history:

- *After the Second World War* , and despite the dreams of Stalin who had built more than 500 submarines, there was a phase of maritime isolationism.

- *From 1955 to the early 1960s* , and in spite of Mr Kruschchev's reluctance, a fleet was slowly built up to carry on trade within COMECON.

- The threat to the Soviet coasts from Polaris, and the humiliation over the Cuba crisis enabled Admiral Gorshkov, *from 1962 onwards* , to put his ideas into practice and the merchant fleet grew in step with the Navy. Progress was spectacular, as borne out in Table 7.1. As the Soviet Navy was becoming increasingly prominent in such theatres as the Caribbean, the Indian Ocean or the Middle East (the Six-Day War), the Soviet shipping companies made their true entrance to the world shipping lanes.

- Finally, *from 1975 to 1985* , this fleet grew more slowly, obviously, and it adjusted to the market, developing its presence in the traffic of OECD countries, particularly with the Far East, Africa and Central America.

The Soviets are today letting it be known to all and sundry that the new 1985/1990 Plan will confine itself to replacing obsolete tonnage, and that overall tonnage will remain stable and even decrease slightly after growing 5.3% between 1980 and 1984.

It should be realised straightaway that the Soviets, after the spectacular growth during the 1960s, are acknowledging that their growth has continued in recent years at a spectacular rate, while most of the Western fleets were shrinking. But the most interesting point to note is undoubtedly the way that the fleet's structure has changed.

iii The structure of the fleet

The Soviets began by carrying out a general rejuvenation of their fleet, which now has a published average age of 13 years, fairly close to that of many Western countries.

Furthermore, the oil-tanker fleet, after expanding in the 1970s, is now falling back to stagnate at about 4.5 million GRT, while the dry-bulk fleet remains in the area of 2 million GRT. This is in striking contrast to the rapid growth in ro-ro tonnage which increased 13% between 1983 and 1985. Container-carriers are developing faster (32% up between 1983 and 1985) with new 'MERKUR'-type vessels (941 TEU).

Hans Jacob Kruse, who was already mentioned, rightly points out that the USSR in the final analysis has a fleet that differs from what its commercial needs would appear to suggest. Contrary to the conventional wisdom, the USSR is not self-sufficient, and has to carry for import, as well as for export, sizeable quantities of raw materials. The most spectacular case is that of grain cargoes: Soviet charterers have an essential role to play in this market which has a structural overcapacity of transport resources, and each year they play that role to perfection. In these circumstances, it is preferable to use foreign vessels since foreign-currency expenditure on them is lower than the earnings of Soviet vessels plying the liner traffic routes, whether connected with Soviet foreign trade or, above all, with cross-trade, as they still only have very little domestic containerised traffic.

iv The place of the Soviet fleet on the markets

There are therefore two separate strategies, corresponding to two types of traffic: bilateral trade on the one hand, and cross-trade activity on the other.

*a) **Bilateral Trade:*** In their official statistics, the Soviets themselves post a score of 54.7% of their maritime foreign trade carried under their national flag in 1984. This is an achievement that would be envied by most Western countries which, in the best instances, are closer to 20 or 25%. This being so, it is not surprising that, in its bilateral relations with the OECD countries, the USSR should carve itself the lion's share of shipping. Japan is perhaps an exception, since it retains 4 vessels in the traffic between Nakhoda and Kobe, but what is to be said of the members of the EEC? The situation everywhere is catastrophic, with as a rule 90% of the traffic carried by Soviet vessels, despite the existence of bilateral agreements which normally provide for balanced ratios. The French experience is exemplary, and table 7.2 highlights this situation, in which the French flag no longer carries more than 8% by volume and 20% by value of bilateral traffic.

For the lines which are normally operated by a joint service, the Soviets have succeeded in moving 60% of the traffic via other ports, mainly under their own flag. Regarding oil, which France imports, the French flag's share is 0.2%, while for cereals, which France exports, its share is 17.4%.

This, the French situation, illustrates the case in point, since all our European neighbours experience the same difficulties. In the bilateral talks, it is easy for the Soviets to mobilise their political authorities at the highest level, since the Merchant Marine issue is viewed from a strategic standpoint. This is not so in Western Europe.

Within the EEC, moreover, there is no serious attempt at exchanging information, no genuine coordination among governments, on these bilateral issues. It is therefore child's play for the Soviets to play off against each other all the conflicts of interest among the different European partners, or to play on their inconsistencies. The example of the deviation of a major proportion of Franco-Soviet traffic via Antwerp is a clear illustration of this.

The Soviets also benefit from the 'lobbying' among European shippers, who defend their freedom to carry at the best price, and to reinforce this tendency, they are developing their control on land, in our countries, over the transport sequence, performing operations which, obviously, we are entirely forbidden to do in the USSR.

As an example, and besides the incentive to CIF purchases and FOB sales by the Europeans, they are forever tightening their control over the freight/forwarding agencies in the EEC in order to channel goods towards the vessels and ports of their choice, Antwerp being their main staging-post for such operations.

b) ***Cross Trade:*** More is known about the increasing share of the USSR in European traffic with third countries, and this also arouses more controversy, since the Soviets are at pains to stress how small their presence is, compared with that of other, much more powerful and aggressive competitors.

A difference from the case of bilateral agreements is that here we must obviously transpose ourselves into the context of a world market for general cargo in which the Soviets are perfectly conversant.

First, studies that cannot be called in question belie this unwanted modesty that the Soviets exhibit when their maritime activity is mentioned. Excluded from American traffic after the occupation of Afghanistan, they are present on 50 of the largest traffic routes, as recalled in the Brostroms study summarised in Table 7.3. 232 ships totalling more than 2 million DWR represent competition that is far from negligible, and from a rapid reading, it can be gathered that the majority of its ships operate in traffic sailing from, or bound for, Western Europe.

Secondly, during the last twenty years, the development of this network has not been a chance occurrence. Besides the main strategic or political motives, the commercial objective of the Soviet lines has been to collect cargo - whether outward or on the return according to case - enabling ships to be filled whose prime mission was related to Soviet foreign policy. Since no reaction has been forthcoming from the OECD countries, the Soviets have been able to fortify their positions on certain important routes, sometimes developing to a marked degree the tonnage on offer. Thus, in the Europe - Far East traffic, by far the largest in which Soviet competition is exercised, the Balt-

Orient Line has increased its slot capacity by 127% compared with 1982.

According to the latest statistics available to us, Balt-Orient is estimated to be carrying 6% of the traffic in both directions between the Far East (excluding Japan) and Northern Europe, compared with 1.1% in 1980. But it should be constantly borne in mind that these data take no account of the activity of the Trans-Siberian railway, which has a crucial role, particularly with respect to Japan. The Soviets have carried on an intense effort to modernise their fleet, steadily substituting "Merkur" type craft (941 TEU) for the Khudozhnik class (780 TEU). They will subsequently be deploying the 1,164 TEU "Saturn" vessels, built in East Germany, thereby largely exceeding their present capacity of approximately 115,000 TEU per year in both directions. With the Trans-Siberian railway (105,000 TEU), the Soviets therefore have a freight capacity of 220,000 TEU, while that of the lines in the five "Hard Core" countries is 550,000 TEU. On the Europe-East Africa routes, the Soviet company controlled 13% of the traffic in 1983, compared with 8.9% in 1981.

However, over and above these crude figures, more of significance can be gained from observing the allegedly "commercial" methods employed by the Soviet competition to penetrate those markets. The shipping companies in market-economy countries seek to carry freight at a rate which ensures that, in the medium term, their investment and operating costs are covered, being market costs - and frequently above the market owing to the social-security and labour-relations constraints that prevail in Western Europe.

For the Soviet lines, the economic logic is quite otherwise: doubtless there is an underlying economic motivation, which is to earn hard currency. But the aim is to maximise such earnings, and not to strike a balance between minimum costs and maximum earnings.

The costs Soviet shipowners enjoy are divorced from market considerations in certain respects:

- vessel purchases are financed by the State under the Plan, at 'domestic' costs, and depreciation is not charged in the way that we are familiar with;

- various state subsidies are given, particularly for bunkering in national or friendly ports;

- social-insurance charges are very low, often because of the very large number of hands brought aboard as conscripts, while the remuneration of career seafarers has nothing in common with world-market practices. Furthermore, comparisons are very difficult to make on account of the massive overvaluation in the official rate of exchange for the rouble. But it seems clear, referring to both labour and holiday conditions, and wages, that the cost of a Soviet crew is below that of the most economic flags. The present cost of a Polish crew on the international chartering market is perhaps a tentative illustration of this: the cost is, on average, more than $2,000 dollars per day below that of a French crew on an average container-carrier, and therefore equivalent to the rate offered by the South-East Asian manpower agencies (comparison based on 1986 data).

Thus, the Soviets will be able to practice one of the most aggressive *commercial policies* , first and foremost in the matter of rates. Their undercutting on Far-Eastern traffic averages 10% and may at times be as high as 30 to 50% for the most lucrative commodities. Table 7.4 gives an example of the undercutting practised by the Baltic Shipping Company in 1984 on the East-bound link between Europe and Taiwan.

A proliferation of such examples could be given for other routes, and these practices have serious consequences in terms of the erosion of the earnings of Conference lines, because they depress rates. Thus, in the Far East, in 1983, the losses of earnings attributable to Soviet undercutting (Baltic Shipping and the Trans-Siberian railway) represented 30% of the total loss of earnings of the Hard Core lines, or more than $30 million. The extra loss of cargoes carried by the Soviets brings the total damage suffered to almost $50 million for 1983.

Obviously, the Soviets do not only play at undercutting. Once they are established in a particular traffic, and are themselves threatened by other highly-aggressive outsiders, their tactics will be to raise their rates to a certain extent, while still remaining below Conference rates. Their gamut of techniques is not in any case confined to just using rates as a weapon: offsetting operations between States, very low prices on land-borne traffic, or the waiving of demurrage charges are terms that shippers can find highly attractive.

Finally, relations with the Conference are not always on the footing of an outsider. The Soviet company may, after practising a 'rate-cutting war', be given 'tolerated' outsider status, or even become a member of the Conference, if the traffic calls for it and the Conference remains strong. In the opposite case, there is no reason for it to tie its hands with a Conference that does not control the market.

The Soviet lines today are members of a dozen or so Conferences (see Table 7.5), often in traffic with LDCs where at times, certain restraints remain on traffic.

It is therefore a most hazardous enterprise to deal with the matter as a whole from the sole standpoint of unfair competition. The goal they pursue appears to be more strategic than commercial. Is it therefore possible to brand competition as unfair that is based on economic principles - and upon their practical application - that differ fundamentally from our own, regarding both finance and the operation of the merchant fleet?

European shipowners are wary of mixing matters, of the intrusion of States in commercial concerns. This is not true of the Soviets. Therefore, in order to develop a strategy of response, we should cast ourselves in the mould of their own way of thinking.

The limits and inadequacies to date in our responses should spur the European countries towards this revolution in people's minds.

3 New responses for Europe

The experience of recent years clearly shows that the responses hitherto made by Europe to the sea challenge of the COMECON countries, and of the USSR in particular, have remained to a large extent without effect.

(i) Present Responses

(a) Bilateral Trade:

The French example can be cited afresh. The maritime agreement signed in 1967 is conventional in form and a number of countries - of every political regime - have signed agreements of this type with the USSR. The text speaks of freedom of navigation and of absence of discrimination. But by virtue of the differences in the 'political, economic and social' systems, the dice are loaded from the very outset. For example, in the USSR, access to ports for foreign vessels is under the authority of the Council of Ministers; River Navigation is reserved to national vessels, while foreign companies are also debarred from port operations and inland routings...

The French shipowners nevertheless succeeded in mounting a combined service in 1969 for shipping lines, which are supposed to call at certain ports in each signatory country on a balanced basis. But nothing has been specified for dry bulk or for oil. At first, the agreement operated quite satisfactorily, and then the French share was seen to drift down in 1972, warranting sharp reactions on the Joint Maritime Commission. This resulted in a recovery in the French share, which subsequently dropped back steadily from 1976 onwards (see Table 7.6). The Soviets had obviously measured the weakness in the resistance of the French authorities and taken advantage of it in a big way.

The drift was halted to some extent for dry bulk (for the sale of grains to the USSR) in 1982, when the French side called forcefully for a balance in shipping. Since that time, the share of French shipping, which had been virtually non-existent, is now running close to 20 per cent. And yet, the steady slide prompted a new and very sharp reaction from the French government in 1985. France demanded of the USSR a re-interpretation of the agreement so that the principle of a balanced sharing of traffic, with monitoring of its implementation, could genuinely be given practical effect.

The Soviets rejected most of the French demands. This is an ironic situation, since the tables are entirely turned. The Soviets are the ones who, because of their system, are pressing for the signature of bilateral agreements. But they ceaselessly invoke the principles of freedom of the sea and of liberalism, accusing their Western partners of seeking to protect an uncompetitive fleet and of hiding behind protectionist cargo-sharing measures.

With the other Western European countries, there is sometimes no agreement, or else agreements exist which allow the Soviet flag to exercise an overwhelming dominance. Is it really possible to imagine that a single country could stand in the way of the unfolding of this strategy?

(b) Multilateral Trade:

In 1979, the EEC instituted a procedure for monitoring Soviet activity in two areas of traffic (Central America and East Africa), and then in 1981 in a third (the Far East). Statistical surveys are solely concerned with tonnages carried and are only communicated with some delay.

In late 1981, the Governments and shipowners of five European countries, Great Britain, West Germany, Belgium, the Netherlands and France, formed the so-called 'Hard Core' Group, designed to notify officially the Soviet authorities and shipowners of the seriousness of the problem. After a week of taut negotiations, several 'General Economic Principles' were approved by the 'Five' and the USSR in Moscow. The agreement recognises the right to cross-trade and to non-discrimination and proclaims the necessity for the governments of the 'Five' to safeguard the existence and viability of their national lines on their national traffic. Hence a number of good-conduct rules were set out, particularly with regard to rates, permanence in traffic or quality of service. Negotiations were set in motion among shipowners, particularly for the Central American traffic (WITASS Conference) where, in the East-bound direction, the Soviets have a dominant position for the shipping of coffee and cotton. Despite the signing of the Moscow agreement, which remained confidential at the time, the discussions among shipowners are marking time, particularly for the Far East, where the Soviets are refusing to include the Trans-Siberian railway (TSR) in the talks. In February 1983, fresh negotiations between the governments, in Hamburg, were inconclusive. But a few months later, and despite the activity of a Finnish company actually controlled by the

Soviets (Scan-Pacific), an agreement was signed between the Soviet shipping company Balt-Orient and the WITASS Conference. No compromise was struck, however, on either the Far East or East Africa where, in increasingly strenuous competitive conditions, the Soviets are not moderating their claims.

After the Soviet government was set a number of deadlines for it to urge its shipowners to take a reasonable position of compromise, it has become clear that by the summer of 1985, first, the situation had worsened in Central America, where the Soviets did not renew the agreement with WITASS, since a conflict had broken out with some South American lines. The agreement only lasted 16 months. Secondly, there was no progress at all on East Africa and thirdly, Balt-Orient broke off all negotiations with FEFC in the Far East, arguing that Soviet external shipping relations were "experiencing new developments".

At the same time, the Russian negotiators also broke off relations with the Far-East-Australia Conference.

At the request of the shipowners of the five Hard Core countries, the governments announced on 19 November 1985 that a monitoring system would be introduced in the coming months on the traffic carried on by Soviet ships on the three aforementioned routes. After intensive consultations among the five government administrations, a further communique, on 28 February 1986, stated that the information required would concern the 'type, tonnage and related freight rates of all cargo loaded in the (Hard Core country) from Soviet vessels trading between the (Hard Core country) and East Africa, Central America and the Far East. The information will be required to be supplied by Russian shipping lines or their agents'.

Obviously, this monitoring exercise is more far-reaching than the one previously instituted by the EEC, which is also continuing to operate. The new system should be far more rapid in operation and will also concern freight rates charged, which has not been without some legal or technical problems in the various countries concerned.

But the shipowners see this decision as only a first step, thanks to the possibility of making an indisputable evaluation of Soviet competitive practices, with particular reference to undercutting which, as they see it, is the most damaging aspect, since it has the effect of depressing rates for the traffic as a whole.

The governments are more hesitant. They have been incredibly slow to react, wasting virtually two years (from June 1983 to June 1985) before genuinely resolving themselves to act. The Soviets unfailingly availed themselves of all that time to convince shippers and probably some government officials as well that their lines - particularly on Far Eastern routes - represented only marginal competition compared with US lines, or Evergreen. It is a fact that both these 'Round-the-World' services, while they inflict significant damage upon the FEFC, are also operating directly against the interests of Balt-Orient. According to Mr Averin, the head of the international division at the Soviet Shipping Ministry, Balt-Orient is only carrying 4% of total tonnage shipped. This figure is worth checking. Moreover, all the signs indicate that the Soviets are modernising their fleet and are seeking to consolidate their share of this vital market. But to these data should in any case be added the results of the Black Sea Shipping Odessa Line (Mediterranean-Far East), and those of POL and DSR which are members of Conferences (the work-sharing again!) and those of the Trans-Siberian railway (which, according to Japanese sources, is reported to have increased its traffic by 12% in 1985 to reach some 100,000 TEU).

By adding these figures together, we doubtless have a clearer view of the true facts of the presence of COMECON carriers on Europe-Far East traffic routes.

Is the Hard Core process exhausted at a time when it is entering a new phase? Have the shipowners overcome - in the case of some - their naivety or their repugnance at governments having a hand in their affairs? Are the States with heavy external trade able to make their maritime interests override those of their shippers, who often mistakenly believe that freight wars would be in their interests? Are the cross-trading States ready to overcome their reluctance to act against other cross-traders?

Mr Averin for his part has not waited for answers to these questions in order to brandish his threats in the Western Press: the EEC, he says, is in the grip of an anti-COMECON feeling which is a symptom of protectionism.

"There may be the need for Soviet authorities at the ports to do something similar [to monitoring] but we must wait to see how EEC monitoring is introduced." (*Lloyds List* , 31 January 1986). This senior Soviet official reminds us that he would rather reach negotiated agreements. This is also the wish of the Europeans, both shipowners and governments, who for several years have ceaselessly voiced the wish for such agreements to come about. But is that really feasible, given the very special nature of the sea challenge from the USSR and the State trading countries?

Can new lines for discussion be opened, mainly at Community level?

ii What is to be done?

The issue raised by Lenin himself in a famous work should be referred to here.

(a) *The problem should be situated at its true political level*

The example of France confirms this. All specialists now acknowledge the emergence of the USSR as a sea super-power. The issue of its merchant fleet, a basic instrument of this strategic power, should be tackled no longer at just the level of the Shipping Ministers, but at that of the Ministers of Foreign Affairs, Prime Ministers or Heads of State. Even after the ousting of Admiral Gorshkov, the political objective of the USSR remains the same, and none of that country's leaders has ever disowned Lenin who, in 1920, wrote in the following terms to his Minister of Foreign Affairs: "The Capitalists and their governments will close their eyes over our activities... They will deliver you the equipment and technical knowledge we lack... They will go to great lengths to dig their own grave".

While not refusing all trading links with the COMECON countries, it is undoubtedly important for us, when negotiating with them, to bear constantly in mind the fact that these immutable principles are there.

Viewing the matter more widely than the mere economic aspect enables us to recall an obvious point, and one on which the Soviets play skilfully in order to encourage a lenient line to be taken in some European circles: the current acute international shipping crisis is basically due to the huge overcapacity in ships, to a large extent generated by an absurd shipbuilding policy, especially in the Far East, with the backing of States, shipyards, shipowners and bankers. Adding to this situation the worsening slide in rates, due as much to the economic situation in certain markets as to the commercially highly aggressive policy - at least for liner traffic - of shipowners in Taiwan and the United States (also subsidised by their governments), there will certainly be a tendency to play down the importance of Soviet competition. Should it be ignored for all that, and should it be allowed to flourish, with all that this can entail for freight rates? Certainly not.

Contrary to their assertions, the Soviets are not ordinary competitors. The freight rates they offer remain as low as they do only for the time it takes to exhaust the majority of their Conference competitors... Then, they have to strike as rewarding a balance as possible, in terms of hard currency, whether within or outside the Conferences.

The shipowners are unable to contain such competition by themselves. It is essential for their efforts to be combined with those of the States, and this must be done more effectively than is the case today. And, above all, with greater determination, since, as Earl Grey reminds us: "Diplomacy without strength is like an orchestra without instruments".

(b) *Where bilateral trade is concerned* , each European country must first of all take the matter to the highest level. No agreement should be maintained with the USSR if it does not provide at the very least for genuine commercial reciprocity, for the shipowners of both contracting countries to have a balanced share, by tonnage and by value, in the carriage of all goods comprised in the trade between the two countries. This must apply equally to bulk traffic and to liner traffic.

Furthermore, the vessels of each country should have access to the rivers of the other (river-maritime traffic) and should enjoy equivalent treatment in the ports.

For freight administration, in the liner-traffic sector, a joint freight agency might ensure fulfillment of the commitments by both parties. It is unreasonable to expect to be able to conduct commercial transactions inside the USSR, but in that case, the activities of Soviet companies in our own countries should be limited, particularly in the field of freight forwarding. Furthermore, it would be normal, in the freight account breakdowns, for Western shipowners to be able to include chartered vessels, and finally for every precaution to be taken so that the Soviets do not impose CIF or FOB terms of trade according to whether European imports or exports are concerned.

This is a vast programme. However, one can see scarcely any other position to adopt on bilateral traffic if one really wished to preserve for our vessels a reasonable share of the traffic. The authorities in Brussels may feel repugnance at such a clear display of the rules of bilateralism, which in other respects they condemn, or only conceive of on a Community footing. Regulation 4055/86 on joint shipping policy dealing with the freedom to provide services scarcely had any inkling of the difficulty of the problem. In shipping, with the USSR and with the other COMECON countries, bilateralism with cargo-sharing is a necessary evil, and owing to the fundamental differences in the political, economic and social systems concerned, any other solution - at least within a humanly-conceivable time-scale - appears doomed to failure.

One may also question what will follow should the bilateral negotiations break down, and the Soviet share continue to grow. The promulgation of laws in most of the EEC countries on the protection of national maritime interests provides in this respect the beginnings of a satisfactory response. Where France is concerned, a system of compensating financial levies or quotas is provided under the Act of 23 December 1983, which cites a whole range of practices or measures prejudicial to France's commercial and maritime interests, and warranting the enforcement of such mechanisms.

A persistent Soviet refusal to grant the French flag a reasonable share in seaborne trade between the two countries certainly falls within this category, if one considers the whole range of restraints to which the USSR subjects its partners preventing genuine and fair competition from being exercised between the two fleets and the operators.

Priority should be given to negotiation. But retaliatory measures should also be held in readiness, even if we do not have to use them. Thorough consultations with shippers, whose interests should obviously be catered for, are an absolute requirement in all circumstances, and provision for this is written into our Act. Similarly, at least for the channeling of information, the reactions of the various member countries of the EEC should be better coordinated. Splendid isolation has never been of much profit in such matters, and there is no certainty that this has been clearly understood by all partners, in spite of the ways opened in Regulation 4058/86 which specifically provides for "coordinated actions to safeguard freedom of access to ocean-going traffic", but which is implicitly aimed more at LDCs than at the State trading countries. Moreover, we should not forget that the ultimate goal of this Regulation is to preserve freedom of access to international sea traffic. This text is therefore aimed against national protection measures taken by States and is designed to maintain, among others, the rights of cross-traders. It is strange that, in all the major international negotiations conducted in the last few years, particularly between the Europeans and the United States on this theme, everyone has considered cross-trade access to the domestic traffic of the COMECON countries to be impossible from the outset.

(c) Concerning *multilateral trade* , there is no doubt at all that the Hard Core exercise should be continued, including as it does only a few members of the EEC.

The object of this exercise is not to bring the Soviets to their knees, nor to change their economic and social system. After an indisputable evaluation of the damage caused - and that is the role of the new monitoring exercise - governments should be made to face up to their responsibilities of either using political pressure to bring the Soviet government officials, and then the Soviet shipowners, back to the negotiating table in order to bring about the "self-restraint" defined in the 1982 Moscow agreement; or taking measures and, in this respect, the laws mentioned earlier on the defence of national maritime interests seem at first sight to be well-suited instruments.

For multilateral affairs, what is involved is unfair competition in European traffic. Hence, we turn towards Regulation 4057/86, which deals with the problem of unfair competition. We are probably facing here one of the paradoxes of the policy, if they ever came to be applied to the Soviet case or to State trading countries. Regulation 4058/86 provides the perfect framework for the action to be taken in common. The type of counter-measures it provides for are entirely in keeping with the goal which could be sought should negotiations fail. But in reality, the case of 'unfair competitive practices', and the very operating principles themselves of Soviet shipowners are virtually a textbook illustration of such practices, and are covered by Regulation 4057/86 on 'unfair pricing practices'.

Obviously, any EEC shipping company could proceed against Soviet shipowners on this basis tomorrow... But those shipowners are specific, and arise from State policy. It must therefore be hoped that after a few court rulings handed down swiftly and effectively, the general behaviour of Soviet shipowners will alter on most of the 'sensitive' traffic routes on which they now operate. This possibility cannot be rejected out of hand, but we are still a long way from the more political process involved in the Hard Core, which is more the province of relations between States. To add to the questions that may arise, one wonders if Regulation 4057/86 should not be enforced as a matter of priority against shipowners who enjoy subsidies and aid of a far less obvious kind than the Soviets, such as some American or Far Eastern shipowners... It is therefore important in all respects to ask at least two questions: Should not the scope of Regulation 4058/86 be widened, with not just certain Third World countries in mind? and should not the scope of Regulation 4057/86 be widened, with in mind not just the State trading countries alone?

4 Conclusions

In concluding this analysis, one can only urge that the issue of maritime relations with the state trading countries should be tackled as dispassionately as possible. No fact adduced concerning the long-term policy pursued by the USSR, and its translation in the maritime field, arises out of an unthinking anti-Soviet attitude. The aim is to refer to the goals of that super-power as that power quite openly acknowledges them. The weakness of some European shipowners is that they do not always treat this issue with the same professionalism they exhibit in their ordinary commercial negotiations. When dealing with this matter, there is an additional and essential political dimension. Governments know this, but are too often hesitant to proclaim the fact openly, and in any case, to make it understood by economic partners who are somewhat inexperienced in these matters. More serious dealings and less rhetoric are doubtless necessary in order to re-situate the responsibilities of each where they actually lie. The world shipping crisis is deep-seated and the very existence of the European fleets, even in liner traffic, is under threat. A common European policy is urgently required to deal with it, and one is glad to see it emerging.

'Normalising' our maritime relations with the USSR and its allies will no doubt contribute something towards a solution to this crisis, since the disruption caused by this super-power is far from negligible, especially on account of its induced effects. However, the true issue is on a far wider front than that of the trade agreements which, it is hoped, will be signed. What is at stake is the determination of our States to contain a sea power which, if we are not careful, could achieve dominance. The mercantile marine is a vital part, an indispensable component of the logistical plan.

Table 7.1: The Soviet Fleet

Total		
1960	1,138	3.5 million GRT
1975	2,278	11.6 million GRT
1985	2,600	16.8 million GRT
of which:		
Tankers		
1960	121	692,000 GRT
1975	481	2,2278,000 GRT
1985	426	4,600,000 GRT
General Cargo		
1960	1,000	2.8 million GRT
1975	1,757	7.3 million GRT
1985	1,7000	7.8 million GRT
Container Ships		
1975	11	61,000 TEU
1985	50	625,000 TEU

Source: Lloyds Register

Table 7.2: Sea Trade Between France and USSR

	Average 1983 - 84			
	Volume	%	Value F.Frs 000's	%
IMPORTS				
French flag	105,000	1.3	1,540,000	9.6
Soviet flag	6,251,000	78.0	11,605,000	72.8
Other flags	1,642,500	20.7	2,807,500	17.6
TOTAL	7,998,500	100.0	15,952,500	100.0
EXPORTS				
French flag	787,500	15.3	4,261,500	32.6
Soviet flag	3,625,500	71.6	7,476,000	57.3
Other flags	676,500	13.1	1,327,500	10.1
TOTAL	5,089,500	100.0	13,065,000	100.0
TOTAL IMPORTS/ EXPORTS				
French flag	892,500	6.7	5,801,500	20.2
Soviet flag	9,876,500	75.5	19,081,000	66.0
Other flags	2,319,000	17.8	4,135,000	13.8
TOTAL	13,088,000	100.0	29,017,500	100.

Source: CCAF

Table 7.3: Soviet Vessels in International Liner Traffic (1.1.1985 to 30.6.1985)

		Total No of vessels	Tons DW	
1:1	North Europe/Canada East Coast	13	19,900	(2)
			19,600	(2)
			19,300	(1)
			19,200	(2)
			12,300	(1)
			11,400	(1)
			11,300	(2)
			4,600	(2)
1:4	North Europe/Central America, Mexico, US Pacific	6	13,700	(5)
			12,500	(1)
1:5	North Europe/Carribean-Mexico-Central and S. America West Coast	18	13,700	(8)
			13,200	(4)
			12,900	(1)
			12,500	(1)
			11,400	(1)
			8,800	(2)
			6,100	(1)
1:6	North Europe/South American East Coast	14	13,700	(3)
			13,200	(7)
			11,400	(2)
			8,800	(2)
1:7	North Europe/Portugal	2	2,200	(2)
1:8	North Europe/North Africa	4	4,200	(1)
			3,100	(1)
			2,200	(2)

(continued)

Table 7.3 (continued)

1:9	North Europe/Eastern Mediterranean	52	14,200	(2)
			12,900	(1)
			12,600	(1)
			9,400	(1)
			9,100	(1)
			7,200	(2)
			6,600	(3)
			6,500	(4)
			6,400	(1)
			5,800	(1)
			5,500	(1)
			4,500	(3)
			4,400	(7)
			4,200	(2)
			4,000	(6)
			3,900	(9)
			3,800	(3)
			3,700	(1)
			3,400	(3)
1:10	North Europe/East Africa	14	13,200	(7)
			12,300	(3)
			11,400	(3)
			8,800	(1)
1:11	North Europe/West Africa	15	7,400	(10)
			4,500	(4)
			4,400	(1)
1:12	North Europe/South East Asia	12	16,000	(5)
			14,500	(7)
1:13	North Europe/Jeddah	same vessels as 1:12		
1:14	North Europe/Iran	20	2,900	(3)
			2,700	(1)
			2,100	(15)
			1,500	(1)
1:15	North Europe/Tilbury (GB)	2	6,300	(2)
1:16	North Europe/Helsinki (Finland)	3	6,500	(1)
			6,300	(2)

(continued)

Table 7.3 (continued)

1:17	Continental Europe/ Middle & Near East & Red Sea	4 (LASH vessels)	8,600 8,400	(2) (2)
		Size and number of barges unknown		
1:18	Continental Europe/ India-Pakistan	1 (LASH Vessel)	38,000	
1:19	Continental Europe/ South East Asia	1 (LASH Vessel)	38,000	
2:1	North Europe and Mediterranean/ Australia-New Zealand	6	21,000 18,500 18,000 17,800	(1) (3) (1) (1)
2:2	North Europe and Mediterranean/ Pakistan-India-Bangladesh	4	13,500 7,700	(1) (3)
2:3	North Europe and Mediterranean/ South East Asia (conventional)	1	13,700	(1)
3:1	Mediterranean/Jeddah	Same vessels as 3:2 and 3:3		
3:2	Mediterranean/Pakistan-India	10	13,500	(10)
3:3	Mediterranean/South East Asia	10	17,700 16,000 15,000 13,500	(2) (2) (1) (5)
3:4	Mediterranean/Iran	10	3,100	(est.)
3:5	Mediterranean (Adriatic)/Egypt	3	7,700	(3)
3:6	Mediterranean (Adriatic)/ Piraeus-Cyprus-Lebanon-Syria	2	8,500 4,300	(1) (1)
3:7	Mediterranean (W.Italy)/Spain	2	9,100	(2)
3:8	Mediterranean/Japan	4	20,100 18,500	(3) (1)

(continued)

Table 7.3 (continued)

4:1	Japan/Hong Kong	Same vessels as 5:1		
4:2	Japan/Manilla	Same vessels as 5:1		
4:3	Japan/South East Asia-India	15	20,000	(1)
			12,300	(1)
			12,000	(1)
			8,300	(1)
			7,400	(9)
			7,300	(1)
			6,300	(1)
5:1	Australia/Japan	6	15,200	(1)
			15,100	(1)
			14,500	(2)
			14,400	(1)
			14,300	(1)
5:2-3	Australia/Hong Kong-Philippines	Same vessels as 5:1		
6:1	Canada Pacific/South East Asia	3	12,300	(1)
			12.000	(2)
6:2	Canada Pacific/Hong Kong	Same vessels as 6:1		
8:1	Hong Kong/Bangkok	Same vessels as 6:1		
8:2	Hong Kong/Manila	Same vessels as 5:1		
8:3	Hong Kong/Singapore-Penang	Same vessels as 1:12 and 6:1		
8:4	Hong Kong/Calcutta	Same vessels as 3:3		
8:5	Hong Kong/Jeddah	Same vessels as 1:12		
9:1	South East Asia/India	Same vessels as 4:3		
9:2	South East Asia/Jeddah	Same vessels as 1:12		
9:3	South East Asia/Bangkok	Same vessels as 4:3		
9:4	South East Asia/Japan	Same vessels as 4:3		
10:1	Philippines/Hong Kong	Same vessels as 5:1		

(continued)

Table 7.3 (continued)

10:2	Philippines/Japan	Same vessels as 5:1		
10:3	Philippines/Singapore	Same vessels as 4:3		
10:4	Philippines/India	Same vessels as 4:3 Service deleted		
10:5	Philippines/Jeddah	Same vessels as 1:12		
TSCS	Japan-Nahodka/Vostochny	8	15,100 14,500 14,300 6,500	(2) (1) (1) (4)
TSCS	Hong Kong-Vostochny	Same vessels as 5:1		
TOTAL		265	2,533,100	
Deduction for vessels mentioned more than once, i.e. operated under several lines		33	401,300	
GRAND TOTAL		232	2,131,800	
Average size of vessel			9,189	

Note: Figures between brackets refer to number of vessels

Source: Brostroms

Table 7:4 Examples of Freight Rates on Routes between Europe and the Far East

Eastbound	Balt-Orient	FEFC
	US $	US $
Aluminium Profiles	1010	2149
Barley & Oats	1028	1303
Beer, Soft Drinks, etc.	1151	1397 (SCQ) Beer/Shandy 1552 Others
Biscuits	1123	1376 (SCQ) Butter Cookies 1528 (SCQ) Others
Electrical Goods	1325	2249
Graphite Electrodes	950	1851
Milk Products	950	1202 (SCQ)Milk Powder Products 1292 (SCQ) Liquid Milk
Provisions	1370	1950
TV Sets	1075	2453
Trucks (Used) & Tractors	1111	1403 Trucks 1652 Tractors
Sanitary Ware	1150 (all-in rate)	1488 (SCQ)
Zinc Products	972	2149

Notes:

(1) All rates are net to which CAF/BAF have to be added.
(2) All are rates in US $ for 20' container
Rates from Germany/Belgium/Netherlands. Rates from other European areas similar or marginally higher.
(SCQ) = Special Commodity Quotation.

Source: CCAF

Table 7:5 Soviet Conference Membership

The Soviet lines are members of the following conferences:

(1) Brazil/Europe/Brazil Freight Conference (128a-c)

(2) Europe Argentina Freight Conference
Argentina Europe Freight Conference (244c-d)

(3) Continent to Australia Conference (139)

(4) Australia to Europe Shipping Conference (121)

(5) New Zealand European Shipping Association (233-234)

(6) Continent West Africa Conference (COWAC)(146/147)

(7) Zürich Agreement (Outwards)

(8) Cyprus Agreement

(9) Continent/India-Pakistan-Bangladesh Eastbound Conference (145)

(10) India, Pakistan/Bangladesh/Continent Westbound Conferences (179-185),(188a/188b)

(11) UK/India Pakistan Bangladesh Eastbound Conference (3)

(12) India Pakistan Bangladesh/UK Westbound Conferences (189/194)

They also have a special arrangement with:

(13) The Association of West India Transatlantic Steamship Lines (WITASS)(24)

Note: Figures between brackets denote pages in Croner's World Directory of Conferences.

Source: Brostroms

Table 7.6: Sea Trade Between French and Soviet Ports (French Share as a percentage of all bilateral traffic)

Year	Percent
1971	21.0
1972	22.7
1973	15.2
1974	22.0
1975	22.2
1976	21.6
1977	16.4
1978	16.0
1979	14.6
1980	14.3
1981	10.4
1982	8.5

Source: CCAF

CHAPTER 8

THE MARITIME POLICIES OF THE NON-EEC SCANDINAVIAN COUNTRIES AND THEIR IMPACT ON EEC SHIPPING

Arnljot Strømme Svendsen

1 Introduction

Of the three Scandinavian countries only Denmark is a member of the EEC. Norway and Sweden stay outside the European Economic Community and will probably do so for the next 5 - 10 years. Norway and Sweden are maritime nations, but their strength as maritime nations has been weakened. As of January 1986 the two countries had together a merchant fleet of 15.6 million GRT or about 20 per cent of the total merchant fleet of the EEC countries, which reached 75.6 million GRT. As of April 1973, just before the shipping crisis started, the aggregate tonnage of Norway and Sweden counted 29.0 million GRT or about 34 per cent of the corresponding tonnage of all the EEC countries that are members today (85.6 million GRT). The numbers referred to here are the tonnage figures for ships under the national flags. We have not included tonnage figures of foreign flag vessels owned by companies in either non-EEC Scandinavian countries or EEC countries. Norway and Sweden have a much less important position *vis-à-vis* the EEC in world shipping today.

As to shipbuilding the position has changed in the same direction. Although we do not have exact figures concerning the shipbuilding capacity, we may take the data for shipbuilding output as an indicator. In 1985 the output of the shipbuilding industries in Norway and Sweden amounted to 13 per cent of the corresponding output from the twelve EEC countries (source: The Motor Ship). In 1973 the share of the two non-EEC Scandinavian countries was 46 per cent of the (twelve) EEC countries of today (source: The Motor Ship).

From a superficial first approximation we may say that the EEC ought not to care about Norway and Sweden in shipping and shipbuilding. They are too weak and probably they will get even weaker in the next few years.

Concerning the merchant fleet Norway has few and rather weak liner companies and its share of the international general cargo trade is very limited. Sweden has a bigger liner fleet, but in the last few years also this sector of the Swedish merchant fleet has lost a number of big vessels. In 1984 the general cargo fleet for Norway and Sweden together amounted to about 9 per cent of their total merchant tonnage. According to OECD statistics (source: Maritime Transport 1984, Paris 1985) the share of the general cargo fleet of the total fleet was 19 per cent for UK, 19 per cent for France, 12 per cent for Italy, 50 per cent for Germany, 39 per cent for the Netherlands etc.

The continual reduction of the Norwegian and the Swedish merchant fleets during the last seven years have been accompanied by a slight increase in their fleets under foreign flags. This slight growth in the fleet of foreign flags has, however, not been sufficient to offset the decline of the tonnage under the national flags.

2 Scenarios for the immediate future

Will the development or the decline of the maritime industries, and in particular the shipping industry, continue in Norway and Sweden in the years to come? Will the relation between the merchant fleet of Norway and Sweden together and the merchant fleet of the twelve EEC countries together continue to develop to the disadvantage of the two non-EEC Scandinavian countries? Are there good reasons for the thesis that the crisis in the maritime industries will also for the next few years hit Norway and Sweden harder than the EEC bloc? Will the non-EEC Scandinavian countries be less and less significant for the competition of cargoes among European shipping companies?

It is always a difficult or hopeless task to make forecasts in economic and business matters when a number of the decisive factors or variables can be fixed and changed by political decisions.

In Norway it is a general opinion among leading shipping men that crises are and will continue to be the normal or the stable element in world shipping. Profits will be - as now - a rarity. Nevertheless there will be possibilities for shipowners and companies with imagination, ideas, clever planning and management. I think, however, that we have to choose between two rather pessimistic scenarios for the next years concerning the maritime industries in Norway and Sweden.

The first case, the worst case and the least possible one implies a continued reduction in the coming years to a very limited capacity, to a minimum for the merchant fleets, the offshore fleets and the shipbuilding capacity. It will probably only be possible to build big tankers and/or bulk carriers at one or two shipyards in Norway/Sweden.

The second, and more optimistic case implies that the capacity of the merchant and offshore fleets will be stabilised at the 1985 level.

3 Maritime policies and shipping

The definition of shipping is not obvious. It is changing. For instance, Norwegian shipping is not only Norwegian shipping companies operating their own vessels under the Norwegian flag with Norwegian seamen and officers onboard. It may also be Norwegian companies operating from Norway or from offices abroad with a number of foreign owned vessels under different foreign flags, where the seamen and the officers come from one or several foreign countries - not from Norway. It can be a number of variations of this extreme prototype. Instead the traditional shipowning, shipmanagement may be more predominating without ownership. A shipping company can do its business through chartered vessels or by leased tonnage.

The internationalisation of the shipping industry in Norway will continue to be the dominating feature in its development. Over 90 per cent of the Norwegian fleet never calls at Norwegian ports. This international orientation is now also seen in the choice of country of registration. To ensure a supply of cargo and capital, even more ships with Norwegian ownership interests are registered abroad (see tables 8.1 and 8.2). Multinational enterprises in shipping lend more freedom to the shipping companies to buy and sell, where and when they want, to choose flags, crews and officers, to attract trades and all new profitable opportunities which arise, and so giving them better chance to survive and to prosper for the benefits of their native country and the shipping industry.

During the past ten years Norwegian shipping companies have built up an extensive fleet of service vessels for oil activities offshore. The shipping companies' offshore activity, ie. both ships and rigs, had a gross turnover of 10,000 million Norwegian Krones (= £1000 million) in their activities on the continental shelves of other countries. The anticipated growth in the offshore activities of shipping companies will have to take place mainly outside the Norwegian continental shelf, i.e. in EEC countries and in all parts of the world.

Offshore service vessels have had a significant growth within the Norwegian maritime industry as well as jack-up drilling rigs, semisubmersible drilling rigs and drillships. Offshore service vessels consist of two important types. Supply vessels transport supplies to fixed oil installations or drilling rigs. A special version is pipe carriers which transport pipes, for the laying of oil or gas pipes on the seabed. Tugboats and anchor-handling tugs deal with the moving of anchors for floating installations like drilling rigs, barges etc., and towing operations. Drilling for oil at sea requires extensive use of underwater services. Diving support vessels and stand-by (or rescue) vessels are other specialised offshore service vessels in addition to seismic investigation vessels, firefighting vessels etc.

So, we may say, that the Norwegian maritime or shipping industry is continuing changing its structure from traditional shipping to:

- Maritime intelligence industry (management, brokerage, engineering).

- Offshore operations of all kinds.

- Aquaculture as an interesting field for further growth.

Seventy per cent of the surface of the globe is the high seas, the utilisation of which requires skills and competence of the highest possible quality. Nowadays the maritime industry is on the way of becoming more and more professional and sophisticated. The industry itself as well as the government, which determines the maritime policies, has decided to transform the traditional activity to a new industry without borders or limits, which can assist the customers in devising solutions and be a constant source of innovation and improvement., That is the philosophy of the Norwegian maritime trader, a supplier of every kind of maritime expertise to an international community. And that is the policy of the Norwegian government.

This philosophy and policy will of course have an impact on traditional EEC shipping and traditional shipping in all other parts of the world.

4 An assessment of the policies

In the two countries the general attitude of the governments and leading politicians regarding shipping has been that shipping is something we have but which we have not to care about. Regarding agriculture the attitude has been quite opposite: agriculture you must care about, protect and subsidise in a number of different ways. By subventions and protection the agriculture industry is slowly being strangled to death. So far, the maritime industries have been alive, because they have had to struggle and compete with competitors all over the world without protectionism and aid from the treasury. The maritime industries, in particular shipping, have in contrast to governmental assistance been curbed by, for example, foreign exchange regulations which have hampered the freedom of action of the shipping companies and delayed the important and necessary process of internationalisation. Concerning foreign exchange regulations Norway and Sweden are worse off than any maritime EEC country. The freedom to sign orders for new ships at foreign shipyards was withdrawn from Norwegian shipping companies in the 1950s. The Norwegian government introduced in the 1970s heavy taxes on new investments, which have prevented or obstructed necessary replacement orders or necessary expansion plans. The transfer of vessels under either Norwegian or Swedish flags to foreign flags has also been forbidden or prevented through the same foreign exchange regulations, although in the last five years there has been a certain liberalisation in Norway.

In shipbuilding the governments of Norway and Sweden paid heavy subsidies to the shipyards in the last 5 - 10 years, but have gradually stopped. Now a number of yards have been laid off and will probably never start shipbuilding again.

Regarding taxation of shipping in Norway-Sweden as well as in the EEC countries there are many and complex systems, regulations and qualifications. The OECD has published an inventory of taxation of shipping in OECD countries (*Maritime Transport 1984, Paris 1985, Chapter 4*) that distinguishes four major categories of fiscal and financial assistance, namely:

a) Grants or loans at reduced rates to the building of ships ordered from domestic shipyards.

b) Operating subsidies for national flag vessels.

c) Tax preference or tax allowances.

d) Balancing of accounts of state shipping lines from the treasury.

In using this inventory of information it should be borne in mind, that differences in fiscal treatment are only a facet of the complete problem which also includes direct and indirect subsidies to shipowners and shipbuilders, ease of access to credit and credit guarantee facilities, differential operating and capital costs. Another consideration is that the special tax provisions described by OECD should be evaluated against the general income tax regime in each country. It is a convincing proof of the complexity of the fiscal legislation concerning shipowners in different countries in the OECD, the majority of which are members of the EEC, that there are twenty-four specifications concerning direct taxation of shipping, with particular emphasis on provisions for depreciation of capital expenditure on ships in arriving at taxable profits.

The regime of taxation, of direct and indirect subsidies to shipowners and shipbuilders, of ease of access to credit and credit guarantee facilities etc is not a constant and lasting regime. There are important changes in many countries almost every year.

Of these two main reasons, the complexity and the continuous change of the regime of taxation, of subsidies and of access to credit guarantee facilities etc, it is a demanding task for research to find which regime is most favourable and has the strongest positive impact on shipping in the country in question and strongest negative impact on shipping in competing countries.

Based on experiences in Norway from the last few years, where a number of Norwegian shipping enterprises have set up subsidiary companies, made take-overs of foreign companies or made joint-ventures in EEC countries as in the Netherlands, the United Kingdom, France etc one can guess that there are significant advantages to harvest for the time being in these countries.

So far, this emerges as one conclusion regarding the effects of the differences in maritime policies between the two non-EEC Scandinavian countries and the EEC countries.

But as indicated before, the situation may rapidly change. In Sweden, for example, a great number of vessels have been transferred to foreign flags in the last years. More recently however, because of some changes in the taxation of seamen and shipping a new tendency is under way: some Finnish owned vessels have been transferred from the Finnish flag to the Swedish.

We can thus draw another conclusion: the maritime policies in the shipping countries of Europe - in particular the fiscal part of the policies - change and will probably change more and more rapidly as a consequence of the shipping crisis. It is, however, hard to say, that the two non-EEC Scandinavian countries will be more eager to change policies than the EEC countries.

5 Liberalism versus market regulation

The Norwegian Shipowners Association wrote in 1986 that the present slump is not only a freight market crisis: governments everywhere are ruining the market mechanisms by regulatory political measures. Thus the international shipping crisis is also a political decision-making crisis. The liberal principle of ruling and organising the framework for industries and business will always be hit by general crises, wars etc.

However, one bright spot is that the first signs of a change from protectionist attitudes toward a growing tendency for governments to let the market have the last word can be detected. As the Association remarked even the most tenacious supporters of governmental controls are beginning to realise that an international subsidising competition among the various state treasuries results in chaos, the very opposite of control. But still we have to accept the OECD investigations, which reveal that shipping relations among OECD member countries (and probably also EEC member countries) are affected by different regimes concerning the application of competition policy to shipping, certain restrictive regimes *vis-à-vis* cargo access as well as different regimes concerning fiscal treatment and financial assistance to shipping.

If we take a look at another OECD report, *Maritime Transport 1983*, we find an alarming analysis of restrictive regimes *vis-à-vis* cargo access, of offshore supply, of liner conferences etc. It is a shameless discrepancy between sacred principles and practices. However, the OECD Maritime Transport Committee may confess that intra-OECD (and probably intra-EEC) shipping relations are not always based on this principle.

Concerning offshore supply, the Code of Liberalisation of Current Invisible Operations for OECD members, does not cover this market. This fact has led to a number of member countries having adopted or accepted a *de facto* discrimination between flags in commercial maritime operations within the exclusive economic zones of member countries. Also Norway has been accused of some discriminatory measures in offshore supply within the Norwegian economic zones, but not by the OECD Maritime Transport Committee. Norway has, however, a big state owned oil company, *Statoil,* the biggest industrial undertaking in the country. It makes a number of contracts for services from supply vessel companies every year. There may be a preference in favour of the national company Statoil, but also to rig owners, because there are links between the board of Statoil and the government in any case through informal channels. As a good principle state monopolies or strong cartels ought to be abolished in Norway and everywhere.

In principle the Norwegian government and the Norwegian shipping industry have welcomed the 1985 Memorandum of the Commission of the European Communities concerning maritime transport policy. The free trade basis of the proposals is highly appreciated. Can we however, be confident that also in the future this liberal attitude will prevail in the EEC? Will the interests of the twelve different member countries of the EEC always be in favour of liberalism and anti-protectionism? In most EEC countries international liner shipping cartel business enjoys total or partial *de jure* exemption from existing restrictive business practices legislation. Even in liberal countries in this respect such as Belgium, Netherlands, Sweden and Norway conferences are allowed to operate without substantial government regulation.

The fight against protectionism in maritime industries can be compared with the ever-lasting fight against sin. Nearly everybody agrees that it ought to be counteracted and abolished but we cannot completely eschew it. And the war goes on.

Norway obviously supports the Commission's analysis of the danger and extent of cargo reservation, and is consequently positive about the idea of introducing formalised procedures for coordinated action.

Also the idea of the Commission of extending the proposed EC arrangements to other OECD countries on a reciprocal basis is supported by Norway and Sweden. Free access to cargoes in ocean trades ought, however, to be applied to all maritime sectors. Together with liner and bulk, activities such as cruise/passenger shipping and maritime offshore activities should be included.

Great Britain has in the last hundred years and more been a noble ally of Norway and Sweden regarding the fight for liberal ideas and practical matters in shipping. Together with some other shipping nations in the EEC with good traditions in free trade policy they can lead the world towards a liberal maritime transport order.

6 Norway-Sweden and EEC shipping: concluding remarks

The EEC shipping market is still very important for the two non-EEC Scandinavian countries. There are many obvious reasons for that. But the maritime industries in the two countries have been weakened heavily in the last ten years, so the relative strength in tonnage and transport capacity etc. has been much reduced as compared with the EEC bloc.

There is little fear in the EEC countries today concerning Norway-Sweden as competitors or as enemies of a liberal regime in international shipping. On the contrary, competitors tend to apreciate in particular strong competitors. So in the long run one hopes that the two Scandinavian countries may help the EEC countries through growing maritime industries, efficient management and tough competition.

But it is a long way to Tipperary!

Table 8.1: The Norwegian Owned Fleet (1.1 1986; 1,000 DWT)

	Norwegian Flag		Foreign Flag	
	No.	Tons	No.	Tons
Tankers	59	7,446	23	1,808
Gas & Chemical Carriers	92	2,024	60	1,274
Combined Carriers	25	2,436	18	1,662
Bulk Carriers	55	3,278	100	4,018
General Cargo etc.	214	1,439	69	694
Passenger Vessels	29	...	8	...
Offshore Vessels	193	283	71	90
Drilling Rigs	18	...	15	...
Accommodation Rigs etc	9	...	3	..
	694	16,906	367	9,546

Table 8.2: Norwegian Owned Foreign Flag Vessels 1.7. 1984

	No.	Contracts
Liberia	77	20
Panama	36	1
U.K.	25	-
Hong Kong	17	5
Singapore	16	4
Brazil	13	-
Canada	13	-
Philippines	12	3
Netherlands	9	-
Colombia	7	-
Sweden	7	-
France	6	-
Other	34	6
	272	39

Table 8.3: The Norwegian Tonnage and Trades (February 1985) (Not Offshore)

	Norwegian Flag 1000dwt.	%	Foreign Flag 1000 dwt.	%
Within Europe	2,161	12	385	5
To and From Europe	9,213	49	3,419	46
Within America	1,107	6	625	8
Other Trades	6,189	33	3,038	41
	18,670	100	7,467	100
Regions to and from				
Norway	1,807	10	353	5
EEC	11,067	59	3,784	51
USA	5,189	28	3,149	42
Brazil	2,024	11	690	9
Japan	4,391	24	1,805	24
Australia	1,357	7	899	12
Arabian Gulf	4,190	22	687	9
	18,670		7,467	

Note: EEC and USA includes trade between EEC and USA

Table 8.4: Norwegian Offshore Service Vessels February 1985. Number of Vessels

	Norwegian Flag	Foreign Flag	Total	%
Norwegian Shelf	91	-	91	42
British Shelf	34	19	53	24
Other Europe	11	6	17	8
America	8	28	36	16
Africa	4	4	8	4
Asia/Australia	2	11	13	6
	150	68	218	100
Vessels unemployed, laid up, repair etc.	37	3	40	
Total	187	71	258	

CHAPTER 9

MARITIME TRANSPORT SERVICES IN THE URUGUAY ROUND

Raymond J Krommenacker

1 Introduction

GATT - the General Agreement on Tariffs and Trade - is a multilateral contract, subscribed to by ninety-four governments which together account for more than four-fifths of world trade. Its basic aim is to liberalise world trade and place it on a secure basis, thereby contributing to economic growth and development and to the welfare of the world's peoples. The General Agreement is the only multilateral instrument that lays down agreed rules for international trade. Since 1948, GATT has also functioned as the principal international body concerned with negotiating the reduction of trade barriers and other measures which distort competition, and with international trade relations. GATT is thus both a code of rules and a forum in which countries can discuss and overcome their trade problems and negotiate to enlarge world trading opportunities. The eightfold growth in the volume of international trade since the Second World War has provided continuing evidence of GATT's success in this double role.

There is no other legal framework that promotes the multilateralisation of trade between countries in all parts of the world, with every kind of economic system and at all stages of economic development. The contractual rights and obligations which GATT embodies have been accepted, voluntarily, in their mutual interest, by the member countries. Overseeing the application of these rules is an important part of GATT's activities. In successive multilateral negotiations in GATT, the rules have been refined and obstacles to trade progressively reduced.

The GATT ministerial level meeting in November 1982 took the following decision:

"The CONTRACTING PARTIES decide:

1 To recommend to each contracting party with an interest in services of different types to undertake, as far as it is able, national examination of the issues in this sector.

2 To invite contracting parties to exchange information on such matters among themselves, *inter alia*, through international organisations such as GATT. The compilation and distribution of such information should be based on as uniform a format as possible,

3 To review the results of these examinations, along with the information and comments provided by relevant international organisations at their 1984 session, and to consider whether any multilateral action in these matters is appropriate and desirable".

Thirteen national studies of issues in the services sector were circulated through GATT in the course of 1984. These comprise studies by Canada, Denmark, the European Economic Community, Finland, the Federal Republic of Germany, Italy, Japan, the Netherlands, Norway, Sweden, Switzerland, the United Kingdom, and the United States. In 1985, three additional national studies were presented, those of Belgium, Australia and France. Currently, a number of developing countries are carrying out similar national studies.

The exchange of information provided for in the 1982 Ministerial Decision took place at meetings open to the participation of. all contracting parties. The secretariat prepared an analytical summary of national studies as well as a summary of information made available by thirteen international organisations. Three of them (International Telecommunication Union, United Nations Conference on Trade and Development, United Nations Centre on Transnational Corporations) were invited to make presentations on specific aspects of their work. The exchange of views at meetings permitted a more thorough analysis of themes and concepts than appeared in the available documentation.

For the time being, the typology that emerges from the analytical summary prepared by GATT in 1985 on the basis of sixteen national studies on services identifies the following ten categories: banking, insurance, transport, telecommunications, business services, tourism, distribution, health, education, and other services not elsewhere specified, for example cultural services. Some categories, such as business services, can be broken down into numerous sub-categories.

Regarding maritime transport, national studies note that traditional regulatory activities in maritime transport involve carrier agreements in the field of rate-fixing, pooling or space charter as well as tariff systems, licensing and certification. Some specific regulations deal with coastal trade or cabotage and require that goods transported by water or by land and water between points in a given country move by vessels built, flagged or registered in the country concerned, documented under its laws and owned by citizens of that country. In a few countries, new regulatory activities extend the cabotage laws to offshore oil-related activities. National security concerns and the promotion of the national merchant marine industry serve as the primary rationale for the application of these regulations. Administrative approval may be required for investment or transfer of national vessels and ship construction facilities. Foreign investment in national maritime transport may be restrained by the ineligibility of foreigners for government subsidies or by limitations imposed on the methods of financing ship construction.

National studies go on by noting that in some countries the primary objective of shipping policy is to ensure a liberal international shipping policy, with some exceptions, under which all ships have access to cargoes in free and fair competition and regardless of the trade flows involved, and shippers are free to make use of the ship-owning companies of their choice. In these countries, foreign ship-owning companies are free to establish and are granted national treatment. In one country, the shipping law has recently been amended to reflect government policy requiring the national shipping line to operate in a commercial manner and allowing it to respond more flexibly to market opportunities and competitive pressures. Another country, which has until recently pursued shipping policies with a minimum governmental interference, has now adopted rules for a balanced market regulation in light of the increasing intervention by various governments.

RESTRICTIONS IN MARITIME TRANSPORT

In their national studies on services, Belgium, Canada, Denmark, Finaland, France, the Netherlands, Norway and the United States have indicated that one or several of the following restrictions may hinder international activities in maritime transport.

1 *Restrictions related to establishment or investment:*

- restriction on the establishment of foreign controlled shipping companies in the form of capital requirements.

2 *Restrictions related to competitive opportunities:*

- cargo sharing, with 40 per cent of cargo reserved for ships of the importing country, 40 per cent for ships of the exporting country and the remaining 20 per cent for other ships; sometimes, cargo sharing on a 50/50 per cent basis;
- 100 per cent cargo reservation when carriage of government goods is involved;
- flag discrimination;
- cabotage legislation reserving part or all of coasting trade and/or offshore activity for national flag ships;
- requirement that local companies handle the sales of foreign shipping companies;
- financial government support (subsidised loans, covering of losses) enabling companies to compete on a non-commercial basis;
- rebate of taxes and fees to the national carrier.

3 *Restrictions related to State trading companies:*

- shipping services offered at prices well below those acceptable to traditional carriers;
- trade only on the basis of CIF selling and FOB purchasing of goods

4 *Restrictions related to standards:*

- unnecessarily strict or unreasonable standards to prevent the use of more efficient equipment by competing shipping companies;
- extensive documentation requirements for shipping companies;
- port entry procedures resulting in long waiting periods before loading or unloading.

In their 1985 meeting, the CONTRACTING PARTIES agreed to convene a preparatory committee, chaired by the Director General of GATT, Arthur Dunkel, to discuss the subject matter, modalities for, and participation in the new multilateral trade negotiations. The Ministers meeting on the occasion of the Special Session of the CONTRACTING PARTIES to the General Agreement at Punta del Este (Uruguay) from 15 to 20 September 1986 adopted a Declaration, Part II of which launches negotiations on trade in services. *The Ministers decided that the negotiations on services would not take place within the legal framework of GATT but that nevertheless GATT procedures and practices would apply to those negotiations.* The objectives of the negotiations are as follows:

> "Negotiations in this area shall aim to establish a multilateral framework of principles and rules for trade in services, including elaboration of possible disciplines for individual sectors, with a view to expansion of such trade under conditions of transparency and progressive liberalisation and as a means of promoting economic growth of all trading partners and the development of developing countries. Such framework shall respect the policy objectives of national laws and regulations applying to services and shall take into account the work of relevant international organisations."

A reading of these objectives shows clearly that the negotiations have a threefold purpose: first, the expansion of trade in services under conditions of transparency and progressive liberalisation; secondly, the promotion of the economic growth of all trading partners; and, lastly, the promotion of the development of developing countries. As regards the ways and means of implementing the results of the negotiations, the Declaration states that it is the Ministers who shall decide at the end of the negotiations. In other words, all options concerning the legal status and institutional framework of an international agreement on services remain open until the ministerial decision is taken.

The negotiators are, one may say, faced with a blank page in that there is no set of principles and rules that can serve as a direct frame of reference. Everything remains to be created on the basis of the three objectives mentioned. The approach, to be defined and applied, will have to maintain an appropriate balance among the objectives of expansion of trade, economic growth and development of developing countries. A multilateral agreement *erga omnes*, i.e. extending to all the trading partners to the negotiations, would have to be the goal of these negotiations. That having been said, it is clear that participation in such an agreement will depend on the type of agreement that can be reached at the end of the negotiations, in particular on the nature of its normative provisions and the degree of progressivity of liberalisation.

By way of analogy, one can point out that the General Agreement was signed on 30 October 1947 by 23 countries which had just negotiated a first reduction in customs duties. These 23 countries included 11 developing countries. It took 39 years for the number of member countries to reach the figure of 94.

2 Attempts at liberalisation

As regards the substance to be included in the multilateral framework, it is interesting to note that at the present time there are certain achievements and many endeavours in the matter of agreement on trade in services that can be referred to by the negotiators of the Group of Negotiations on Services. There is, first of all, the extreme variety of the existing regulations and of the planned regulations in the

European Economic Community (in maritime transport as well as in the other services), in particular in the context of the process of establishing a single internal market by 1992.

Along with these Community activities of a regional nature, which are supplemented by a great number of reforms at the national level, such as the French legislation on telematics, (which would represent a major change to the law signed by King Louis-Philippe on 2 May 1837, institutionalising a telecommunications monopoly, whose provisions have inspired numerous foreign legislations), there exist two bilateral agreements.

First, there is the agreement in force between the United States and Israel, of a non-binding nature, under which detailed discussions are being held on three services (tourism, insurance and telecommunications) in order to test the practices in these three fields against the principles enunciated in the agreement and to identify the areas in which there may be inconsistency. The principles include national treatment in the application of regulations affecting services, fair play by monopolies in the purchase and sale of services outside the area of their reserved legal monopoly, and full transparency in domestic law and regulations affecting trade in services. Regarding services regulated by a political sub-division, the Declaration provides for the central government to consult on a "best efforts" basis with its political sub-divisions to assure that regulations of the bilateral sub-division are consistent with the principles of the Declaration. The principles also provide for a guideline for a review process in which the United States and Israel are to consider how the current services declaration may be transformed into a legally binding agreement (paragraph 9). Issues relating to professional services were first addressed with regard to extending national treatment to United States travel agents, i.e. the possibility for them to enter into a contractual relationship with a licensed Israeli travel agent who then would assume the legal responsibility for administering the provision of travel services to United States tourists in Israel.

The second bilateral agreement is one that is being negotiated between the United States and Canada. Telecommunications and information services received priority attention in the Canada-United States Free Trade Area negotiations. Several principles being examined for the bilateral free trade agreement are reviewed to see how they may be applied to these services. It seems that a better understanding of each side's regulatory environment is needed in order to advance the process of working out an agreement. This concerns for example the definition of basic and enhanced services for domestic or international purposes, the policies on the resale and shared use of telecommunications services, the availability of value-added and information services by use of private leased circuits, the possibility for private firms to compete for enhanced services with telephone monopolies, the pricing policies for telecommunications transport services and the co-ordination between provincial and federal regulations.

Another bilateral agreement on a particular service which needs to be mentioned is reflected in a law of the Japanese Diet implemented on 1 April 1987. The law entitled 'Special Law on the Handling of Legal Matters by Foreign Lawyers' incorporates the results of the negotiations held between the United States and Japan. The key provisions include: (1) the licensing of foreign lawyers as foreign legal consultants, and the qualifications for obtaining such a licence; (2) the authority of the Japanese Federation of Bar Associations in supervisory and disciplinary actions involving foreign lawyers; (3) the scope of practice, i.e. the kind of services which can be provided; (4) the right of foreign legal consultants in Japan and law firms abroad to share offices and expenses even if the bill does not allow the forming of partnerships with Japanese lawyers, and (5) the right to practice under the internationally recognised name of the foreign firm.

The national, bilateral and regional activities that have just been described demonstrate, if that were still necessary, the importance of liberalisation issues in the trade in services between certain trading partners because of historical reasons or long-established preferential economic relations. *This state of affairs might be categorised by the term 'interest-lateralism' to describe those processes in which shared interests motivate economic policy actions of a national, or a bilateral, or, lastly, of a plurilateral nature.* A particularly interesting aspect of this *interest-lateralism* as regards the Uruguay Round negotiations on trade in services is that it quite naturally provides a kind of full-scale laboratory in which principles and rules can be tested to verify their applicability to the multilateral framework for services under negotiation.

3 The framework programme

A programme for the initial phase of the negotiations serves as a frame of reference for the discussions. The programme consists of the following five elements:

(1) Definitional and statistical issues;

(2) Broad concepts on which principles and rules for trade in services, including possible disciplines for individual sectors, might be based;

(3) Coverage of the multilateral framework for trade in services;

(4) Existing international disciplines and arrangements;

(5) Measures and practices contributing to or limiting the expansion of trade in services, including specifically any barriers perceived by individual participants, to which the conditions of transparency and progressive liberalisation might be applicable.

Two of these elements [i.e. (1) and (3)] should make it possible to sketch the outlines of a possible agreement in what one can call its vertical and horizontal dimensions. The vertical dimension concerns the *definition of trade in services* , namely whether or not it means going beyond the traditional notion of transborder trade and how far. The horizontal dimension relates to the *coverage of the multilateral framework*, namely what types of services should be covered by the multilateral agreement and by possible sectoral disciplines.

In what follows, the chapter dwells in greater detail on the first of these two elements since the parameters of element (3) may well emerge from the negotiations as they approach their final phase.

The typology of services as such does not pose special problems, except as regards the uncertainties inherent in such an exercise owing to the continual emergence of new types of services resulting both from the merging of traditional services and from the impact on services of the application of technological developments particularly in the spheres of telematics and space. Classical examples of these new types of services are direct broadcasting by satellite and value-added networks. However, even this last example is now becoming somewhat obsolete to the extent that the infrastructures (or 'infostructures') of networks (also denominated N2) are becoming software and information intensive. In other words, the usual dichotomy between basic telematics infrastructures and value-added networks has to be adapted to the new technological reality, since it is increasingly difficult to find out where value-added is created in a network. In addition, the current vision of a horizontally stratified telematics marketplace, animated by a model of a network as a pyramid, will not survive. In the last 10 years, an inversion in the relative costs of switching versus transmission has transformed the network architecture from that of a pyramid to a ring-like 'geodesic' where switches and other intelligent nodes proliferate and transmission is kept to a minimum. As a result of this 'geodesic' network, one of the major telematics challenges in the next decade will be the seamless interconnection of the intelligent nodes and thus the permanent and generalised creation of value-added.

As regards conceptual categorisation of types of international transactions in services, four main categories can be identified that take into account the mobility of consumers and of the factors of production: across-the-border trade; domestic-establishment trade; foreign-earnings trade; and third-country trade. 'Across-the-border' trade is the category similar to merchandise trade, in which producer and consumer are each in different countries. In the other three categories, producer and consumer are in the same country but at least one is a non-resident. 'Domestic-establishment' trade assumes that the consumer will move to the producer. 'Foreign-earnings' trade assumes that the producer must have a presence in the consumer's country in order effectively to deliver the service, while 'third-country' trade assumes that both consumer and producer are in a foreign country when the service is delivered.

Lastly, it must be noted that in order to arrive at maximum theoretical and practical consistency, one might envisage using a 'broad' definition of international transactions in services that would include all forms of sales involving a transfer of ownership between nationals (or firms under their control) of a given country and nations (or firms under their control) of another country. The definition of 'ownership' would mean that international trade is considered to take place if the firms or consumers involved are of different nationalities regardless of the location: thus, if a French firm in Hong Kong purchased consulting services from a Belgian firm also in Hong Kong, the transaction would be counted as an international trade in services. That would represent a break with tradition. One could mention that the classical 'location' definition means that international trade is considered to take place if the firms or consumers are in different nations regardless of nationality; thus, if a French firm in Hong Kong purchased consulting services from a French firm in Bangkok, the transaction would be counted as an international trade in services. The fact is that the traditional instruments of economic analysis were not designed to apply to services, which cannot be sold without being simultaneously produced, and vice versa.

Another element of the programme for the initial phase of the negotiations concerns *'broad concepts on which principles and rules for trade in services, including possible disciplines for individual sectors, might be based'*. Under this heading, the object is to examine and develop appropriate concepts that take fully into account the specific nature of services but also the various concerns mentioned above. In particular, this will involve establishing an appropriate balance between the objectives of expansion of trade in services and the objectives of development of developing countries, and of economic growth of all trading partners. Some concepts and issues have already been studied in part, such as transparency, non-discrimination, national treatment, market access, treatment of monopolies, suitable character of a regulation, exceptions and safeguards, application of the agreement in the case of federated States, infant industry and restrictive business practices. These concepts, however, relate to only one aspect, and it is obvious that the negotiations will have to produce principles and rules relating to a larger range of concerns. In the present initial stage of negotiations reference is also made to other possible concepts such as 'progressive liberalisation', 'mutual advantage for countries', and 'economic development'.

Substantial discussions are still needed to deepen the meaning and to test the applicability of these concepts. One would prefer to refrain at this time from suggesting how these concepts might be defined and applied. *It goes without saying that this task belongs to the countries participating in the negotiations.* Nevertheless, it would be useful to refer briefly to seven important issues.

4 Key issues

The first important issue is *national treatment* . This is a complicated question in the case of services because it touches on the operations of foreign service firms (or their dependencies) in the markets of purchasing countries, in addition to the local treatment of services sold across borders. Therefore, before agreement can be reached on a national treatment principle applicable to services, it must be clear whether and to what extent services traded on an investment/establishment basis will be covered in the negotiations. Moreover, it may be necessary to distinguish between national treatment and equivalent treatment, given the fact that in certain circumstances the sheer granting of equal operating rights to dependencies of foreign firms is not sufficient to equalise conditions of competition. In such cases, national treatment may be a necessary, but not a sufficient condition for effective market access.

In the course of the negotiations, the point has already been made that, since the introduction of national treatment principle in the multilateral framework of trade in services would mean full elimination of protection, such an introduction could only be a long-term goal. Consequently, this principle should not be included as an obligation in the multilateral framework. Instead, this principle should be used as a criterion to judge whether particular perceived obstacles to trade in services should be eliminated or not.

The second issue relates to State intervention in services production and trade. This has so far arisen mainly in connection with telecommunications *monopolies*. In this specific context, the question is how to draw the dividing line between basic services (subject to regulations limiting or excluding competition) from value-added networks (open to domestic and foreign competition) since it may appear that services which are basic from one point of view are added from the other and vice versa in the context of the networks generated by the emerging global information economy. But State intervention also has wider implications for the development policies of certain countries where other services activities such as banking, insurance and some types of transport are nationalised. This issue is related to the issue of motivations for regulations, and as in that case, the problem to be solved through negotiation is: where does one draw the line between measures which restrict market access and measures taken for 'other' purposes? In other words, is there any scope for deregulation and how does one define the parameters of the process?

Third, negotiations would also need to address the issue of labour *migration*, in two ways: first, the possibility of firms operating abroad temporarily to employ staff from their home country to perform specific service functions and/or carry out specific service projects, for example construction contracts. Second, broader issues of labour movements may also need to be tackled, at the request of labour-exporting countries. Here, the problem would be to distinguish between primary factor movements which are unrelated to specific service outputs and those which are, and different solutions may be called for in either case.

Fourth, the *treaty-making power* of central governments deserves particular attention in so far as in certain federal states the regulation of major services sectors such as banking and insurance falls within the competence of political sub-divisions. This is not a new problem in the GATT, but it has not been handled consistently: Article XXIV:12 of the General Agreement, Article 7 of the Subsidies Code and Articles 3, 6 and 8 of the Standards Code may confer different degrees of responsibility on central governments for measures taken by local governments; up to now, however, the CONTRACTING PARTIES have not formally given an interpretation of the scope of these provisions. In any event, it does not seem conceivable that services sectors falling under local jurisdiction in certain countries could be excluded from negotiations on that account alone.

Fifth, *development-related issues in services* is another issue to come up in the negotiations. The need for special treatment or appropriate escape clause would appear to depend on the type of arrangement concluded. But besides this well-known approach, an alternative one may be to build on the premise that while all trade expansion may not further development, much of it can do. The challenge for negotiations will therefore be to identify the conditions under which the services trade expansion promotes development, due account being taken of the positive and negative interlinkages between trade and development as well as the different situations from one services sector to another and in the stages of development reached in a country.

Sixth, one major issue, already referred to above, is the existence or not of a *trade-off* between trade concessions on goods and trade concessions on services. This problem not only relates to the eventual outcome, but also to the conduct of negotiations, and it has therefore as much to do with modalities as with objectives. Some governments are of the opinion that some degree of trade-off would be involved, at least implicitly, and for individual governments the question is of importance from the point of view of the kind of support which they can master from the private sector.

Seventh, with regard to *transparency and dispute settlement*, the point is that GATT and other international and regional fora have had a long tradition in the treatment of notifications, consultations, surveillance and dispute settlement procedures of trade in goods and economic policies. In the context of trade in services, it should be noted that in addition to the obligations contained in the OECD Codes concerning various services, the 1985 OECD Declaration on Transborder Data Flows seeks transparency in regulations and policies relating to information, computer and communications services affecting transborder data flows. The governments of OECD member countries have agreed to co-operate and to consult with each other in carrying out this work.

Furthermore, it should also be noted that the United States/Israel Declaration on Trade in Services (L/6019) provides for three complementary stages of transparency and review. First, each party will make public its domestic laws and regulations affecting trade in services and notify the other party of laws and regulations which discriminate against a service exported from the other nation. Each party will provide to the nationals and companies of the other nations reasonable access to established domestic review and judicial proceedings relative to regulations on trade in services. Second, each party agrees to consult with the other periodically to discuss specific problems that arise concerning trade in services between the two nations and to review existing regulatory regimes of the two parties as they affect trade in services. Third, the parties will review the effectiveness of the Declaration not later than eighteen months from the date that it is signed. In this review, the parties will explore further opportunities to strengthen open trade in services between the two nations, including the possibility of transforming the provisions of the Declaration into legally binding rights and obligations.

Existing practices of transparency and dispute settlement in GATT show that this notion is inseparable from those of notification, consultation and surveillance. The question is whether the existing practices could be transposed *mutatis mutandis* from trade in goods to trade in services with respect to *a priori* notifications, the possibility of commenting proposed laws and regulations, reverse notifications, periodic multilateral reviews of reservations and derogations, domestic reviews and juridical proceedings, and the setting up of national Services Enquiry Points (SEP) with the mandate of answering all reasonable enquiries regarding any adopted or proposed laws and regulations governing specific services and services transactions.

The fourth element of the programme for the initial phase of the negotiations concerns *existing international disciplines and arrangements*, for there already do exist, in such services sectors as *civil aviation*, *maritime transport* and *telecommunications*, international agreements which govern certain aspects of trade in services. Consequently, the task is to examine the relationship of these international agreements with the objectives of the negotiations on services. For instance, in maritime transport services, at least two international disciplines have directly relevance to this fourth element, namely the UNCTAD Code of Conduct for Liner Conferences and the 1965 Convention on Facilitation of International Maritime Traffic.

Lastly, as regards the final element, which deals with *government measures and the practices of private firms*, including specific barriers contributing to or limiting the expansion of trade in services, one object is to examine the relationship between the regulatory frameworks existing in the various countries and the objective of expansion of trade as construed in the Ministerial Declaration. In other words, it seems necessary to reach agreement on how to deal with the question of national regulations concerning services and their effects on trade in services, in as much as this element will play a central and decisive role in any multilateral agreement. Another major issue is the examination of how such measures and practices are promoting trade.

5 Concluding comments

The above considerations show what a complex task the negotiators are engaged in, especially as ten major services sectors, and consequently a multitude of types of services transactions, have been identified so far. Nevertheless, there is reason to believe that, in the context of the global market that is emerging, the present negotiations will determine the exact nature of the multilateral trading system governing trade in services for decades to come and of the issues resulting therefrom for the trading partners in various categories of services and, among others, maritime transport services.

The bulk of international services transactions is now taking place in a legal vacuum, without any link between one set of national regulations and another. Trade flows can suddenly be interrupted at the whim of a ministerial or legislative act of any country, and the absence of common disciplines means that only bilateral negotiations, sometimes influenced by arbitrary considerations, can restore them.

Because of the globalisation of markets, there is a growing interest in bringing about consistency between the regulations of different countries, in order to ensure a continuation of a process that promises a great deal of growth and well-being. To move progressively from a context of *interest-lateralism* to an interwoven network of rights and obligations, symbolic of multilateralism, is the challenge being thrown down to businessmen, scholars and government representatives interested in promoting the expansion of trade in services.

INDEX